Infosys Science Foundation Series

AF805899

Infosys Science Foundation Series in Mathematical Sciences

The **Infosys Science Foundation Series in Mathematical Sciences,** a Scopus-indexed book series, focuses on high-quality content in the domain of pure and applied mathematics, biomathematics, financial mathematics, operations research and theoretical computer science. With this series, Springer hopes to provide readers with monographs, textbooks, edited volume, handbooks and professional books of the highest academic quality on current topics in relevant disciplines. Literature in this series will appeal to a wide audience of researchers, students, educators and professionals across the world.

Dia Zeidan · Arturo Hidalgo · Lucy T. Zhang ·
Eric Goncalves Da Silva
Editors

Computational Fluid Dynamics: Novel Numerical and Computational Approaches

Methodology and Numerics

Springer

Editors
Dia Zeidan
School of Electrical Engineering
and Information Technology
German Jordanian University
Amman, Jordan

Lucy T. Zhang
Department of Mechanical
Aerospace
and Nuclear Engineering
Rensselaer Polytechnic Institute
Troy, NY, USA

Arturo Hidalgo
E.T.S.I. Minas y Energia
Departamento de Ingeniería Geológica y Minera
Center for Computational Simulation
Universidad Politecnica de Madrid
Madrid, Spain

Eric Goncalves Da Silva
Department of Fluid Mechanics
and Aerodynamics
ISAE-ENSMA, Institut Pprime
Chasseneuil-du-Poitou, France

ISSN 2363-6149 ISSN 2363-6157 (electronic)
Infosys Science Foundation Series
ISSN 2364-4036 ISSN 2364-4044 (electronic)
Infosys Science Foundation Series in Mathematical Sciences
ISBN 978-981-97-8151-5 ISBN 978-981-97-8152-2 (eBook)
https://doi.org/10.1007/978-981-97-8152-2

Mathematics Subject Classification: 76A02, 76B03, 76D03, 76M12, 65M08, 35S99, 68W99, 65Z05

This Springer imprint is published by the registered company Springer Nature Singapore Pte Ltd.
The registered company address is: 152 Beach Road, #21-01/04 Gateway East, Singapore 189721, Singapore

Preface

Significant development of computational fluid dynamics (CFD) has been made over the past few decades. Major topics of CFD are theory, computation and numerical simulation of different real-world applications. Such topics are based on either scientific methods or experimental design. The purpose of this book is to present novel numerical and computational approaches of various real-world applications using CFD. It is, therefore, of primary importance to invite international expert researchers dealing with CFD to submit their latest research outcomes in the field for possible publication in this book.

CFD consists of a different collection of topics that are distributed among a variety of conference proceedings, journals and books. These are indeed accurate and offers possible tools for many learners in the world. The present book brings another opportunity in the simulations and applications of CFD to the attention of those learners. Consequently, this book is an effort towards presenting the main important topics in CFD and their applications in science and engineering.

Organized into seven chapters dealing with compressible and incompressible fluid flows, heat and mass transfer problems, each chapter of the book discusses one unique CFD phenomena associated with fluid flows. It starts with the introduction of the physical model and the underline mechanism of CFD phenomenon. It follows by the systematic theoretical investigations, numerical and experimental verification of CFD applications when possible. Chapter "Solid Particle Movement Simulation in a Quiescent Container Based on Discrete Element Method Using CFDEM" presents a series of numerical simulations by using CFD-DEM to investigate the vertical movement of microplastics in quiescent water. In Chapter "A Central Finite Volume Surface Gradient Method for the Two-Layer Shallow Water Equations with Rigid Lid", authors have developed a new central finite volume surface gradient for approximating the two-layer shallow water equations with rigid-lid and variable bottom topography. Chapter "Compressible Enstrophy Transport for Flow in a Low-Pressure Turbine with Unsteady Wakes Impinging at the Inflow" investigates the two-dimensional compressible Navier–Stokes equations by using a dispersion relation preserving methodology. Chapter "Improving the Cooling Efficiency of Microchips by Vapor Chamber" deals with CFD simulation for improving the cooling efficiency of microchips in a vapor chamber. Authors of Chapter "A Comparative Study of Correlation-Based Transition Models for Aerodynamics in CFD Code NSMB" present the formulation and implementation of various correlation-based transition models within the framework of an in-house CFD solver, namely NSMB. Chapter "Exact Riemann Solution for a Homogeneous Equilibrium Model with Four Equations" develops an interesting theoretical and numerical solution for two-phase flow equations on the basis of the Riemann problem. Finally, Chapter "A Comparison of $k - \varepsilon$—Type Turbulence Models for Prediction of Inclined Dense Jets" investigates and evaluates the performance of four turbulence models for the simulation of an inclined dense jet in a stagnant water environment.

This book targets scientists and students at all levels who would like to work with CFD problems within different fields of science and engineering—with the hope to many new interdisciplinary lines of CFD research.

Shamim Ahmad, Executive Editor at Springer Nature, has our sincere gratitude for all of his help and encouragement during the production of this book. Additionally, we would like to express our gratitude to all of the authors for their outstanding contributions to this book as well as to the anonymous reviewers who assessed the chapters and offered advice and commentary. We also express our gratitude to the German Jordanian University, Jordan, Universidad Politécnica de Madrid, Spain, Rensselaer Polytechnic Institute, USA, ISAE-ENSMA, Institut Pprime, France for their kind support in helping us publish this book and encouraging novel numerical and computational approaches of CFD research.

Amman, Jordan — Dia Zeidan
Madrid, Spain — Arturo Hidalgo
Troy, USA — Lucy T. Zhang
Chasseneuil-du-Poitou, France — Eric Goncalves Da Silva

Contents

Solid Particle Movement Simulation in a Quiescent Container Based on Discrete Element Method Using CFDEM

Seyyedmahdi Ostadhosseyni, Abdolmajid Mohammadian, and Abolghasem Pilechi

Abstract There has been an increased interest over the past few decades from the engineering and scientific community regarding the transportation of microplastics in open water environments. As a result of the number of processes involved in the movement and behavior of microplastics in water, as well as their significant impact on aquatic environments, the movement and behavior of microplastics in water represents a complex problem. In order to describe the behavior of particles in water, one of the most important steps is to obtain the sedimentation (or floating) velocity, also known as the terminal velocity. With the use of Computational Fluid Dynamics (CFD) coupled with Discrete Element Methods (CFD-DEM), series of numerical simulations have been conducted to investigate the vertical movement of microplastics (settling or floating) in quiescent water. Several simulations were conducted in order to evaluate the potential capabilities of the CFD-DEM method to predict the behavior of microplastics in water under various conditions. A series of simulations of the terminal velocity of microplastics that are represented by round particles has been conducted by using different types of microplastics that are represented by round particles ranging in size from 1 to 5 mm. It was found that the simulation results and experimental data were in good agreement, showing that CFD-DEM can be used as a viable approach for predicting microplastics fate and transport under a wide range of environmental conditions.

S. Ostadhosseyni (✉) · A. Mohammadian · A. Pilechi
Department of Civil Engineering, University of Ottawa, Ottawa, ON K1N 6N5, Canada
e-mail: sosta032@uottawa.ca

A. Mohammadian
e-mail: amohamma@uottawa.ca

A. Pilechi
e-mail: abolghasem.pilechi@nrc-cnrc.gc.ca

A. Pilechi
Ocean, Coastal and River Engineering Research Centre, National Research Council Canada, Ottawa, ON K1A 0R6, Canada

D. Zeidan et al. (eds.), *Computational Fluid Dynamics: Novel Numerical and Computational Approaches*, Infosys Science Foundation Series,
https://doi.org/10.1007/978-981-97-8152-2_1

Keywords Terminal velocity · Particle sedimentation · Particle floating · CFDEM

1 Introduction

The rapid increase in the number of microplastic particles found in the environment over the last few decades has created a growing need for a better understanding of the behavior of these particles in water. Particle behavior has been previously studied in a number of studies, a few of which will be discussed in the following paragraphs.

These studies focused on factors affecting the particle's behavior in air, including drag coefficients and Bagheri and Bonadonna [1] reviewed the most recent methods for describing particle size and shape, which are essential for sedimentation and particle dispersion models. It discussed how to quantify the size, shape, and aerodynamics of regular and irregular particles in order to determine particle settlement velocity. This was done with the least amount of error possible. Saxby et al. [2] compared predictions and measurements of fall velocities in air for different shapes of volcanic ash in order to evaluate the effectiveness of drag equations and shape descriptors in simulating the sedimentation process. In order to obtain more accurate results, shape-dependent drag equations should be taken into account. Fan et al. [3] developed two separate drag coefficients for super-ellipsoidal particles by simultaneously measuring the windward, or projected surface area of the particle normal to its motion direction, surface area and the terminal velocity of the particles during settling experiments. This study led to the development of a more accurate drag coefficient that can be applied to both spherical and super-elliptical particles.

Furthermore, several studies have examined the influence of particle shapes on their movement in water. Yokojima et al. [4] investigated the effect of particle form on the settling performance of sand elements in a quiescent water environment by using a two-dimensional high-resolution numerical model. The presence of adjacent particles obscures the influence of particle form on vertical movement. Furthermore, the terminal settling velocity of the round particles was significantly different from that of the other particles. Based on the findings of this study, it is recommended that the assumption that particles are rounded be used with caution. In addition, it is recommended that further investigation be conducted into how particle shape impacts particle behavior. Zhou et al. [5] demonstrated how particle shape, size, and density have a significant impact on terminal velocity by studying the sedimentation of various disc-shaped particles. The study identified particle form factors that should be taken into account when modeling drag coefficients.

Chen et al. [6] looked into how particle size, shape, and surface roughness affected the speed of sedimentation. They used an image trace approach to measure the sedimentation velocity and the geometric properties of the particles for irregular-shaped particles. The Stokes equation was modified by a correction coefficient and identified a linear relationship between the correction coefficient and equivalent diameter. Karimnejad et al. [7] investigated the geometry and size effects of falling elliptical particles using a hybrid direct forcing Immersed Boundary-Lattice Boltzmann method, which is an explicit scheme acting as a numerical particle-based method. For a single ellipse landing in a thin box, the impacts of aspect ratio, density ratio, and beginning orientation as well as different configurations of the dynamics of two similar ellipses were examined. 20 particles of various sizes and shapes were subjected to simultaneous sedimentation to prove the effectiveness of the proposed technique. It was found that larger particles switch places with smaller ones, while smaller elliptical particles, in contrast to larger ones, tend to hold onto their initial orientations. Zhou et al. [8] has constructed a generalized and practical model to determine the terminal velocity of particles (cuttings and proppants) during the settling process in supercritical carbon dioxide (SC-CO2) drilling and completion. The two main forces acting on particles during the settling process are effective gravity and drag forces. The settling of particles is discussed, and tests are done on terminal velocity to see how different parameters affect the terminal velocity of particles. The terminal velocity is directly proportional to temperature, particle diameter, and particle density, while it is inversely proportional to pressure since under the identical circumstances, there is good agreement between the experimentally measured and numerically estimated terminal velocities.

Hwang et al. [9] used the CFD-DEM coupling technique to model solid particles sedimentation in a water pool. They compared the simulation results to the scaled-condition experiment. The study focused on particle falling parameters (average particle velocity and particle falling time). The range for the particle density is from 3600 to 7800 kg/m^3, which means the study was just about the falling of particles. The study found the CFD-DEM approach as a suitable tool for simulating particle sedimentation on a local level.

Xie et al. [10] used a coarse-grained CFD-DEM approach in which larger particles were scaled up from smaller particles in order to address the problem of high computational costs. Considering different systems of particles such as coarse-grained and original-sized, the effectiveness of the coarse-grained DEM model has been verified in terms of qualitative evaluations and quantitative comparisons. Isachenko [11] used a straightforward stochastic numerical model to determine the distribution of the terminal settling/rising velocity of a group of MPs particles according to production patterns and field observations. It is demonstrated that the simulation results precisely correspond to the data from laboratory tests. Compared to the effect of the distribution by form, the distribution of MPs by size and density seems to be more helpful for the final terminal velocity distribution. Terminal velocity is dependent on multiple physical variables at the same time.

Huang et al. [12] studied the movement patterns of particles, the long-term sedimentation of multiple circular particles in a narrow channel filled with a viscous fluid using numerical methods. They considered two main focuses: the equilibrium motion scheme of particles at a middle terminal Reynolds number (Ret) and the connection between the scheme and the behaviors of multiple particles. They demonstrated that one or two particles are enough to be studied regarding movement patterns in order to obtain key information about multiple particle self-organization behavior. In terms of spatial discretization, analysis of flow across bodies with complicated surface geometry can be difficult. Kalman [13] investigated how the free and tapped void fractions were affected by particle form. High circularities were thought to be equivalent to sphericity based on the examination of well-defined geometries. By dividing the non-spherical particle void fractions by the spherical particle void fraction, the shape factor and Archimedes number (Ar) factor can be determined. Also, Fonceca et al. [14] considered an explicit approach to the body force acting in the fluid phase while calculating the local momentum balance equations. They figured out how to make immersed particle motion in a fluid phase. The resolved CFDEM coupling framework (which will be defined in Sect. 2.3) was employed to solve the particle–fluid interaction. The new implementation precisely reproduces Stokesian dynamics (a method for solving the Langevin equation, which represents the appropriate application of Newton's second law to a Brownian particle).

Hazzab et al. [15] has investigated the sedimentation of simple and complicated solid particles experimentally. The applied data for isometric particles and spherical particles of the same volume have been compared. This comparison makes it possible to highlight the idea of an equivalent sedimentation diameter as well as to make a computational model for the direct calculation. This model performs well when compared to the outcomes produced by other models when applied to the data of the measures under consideration. Tassew et al. [16] investigated utilizing microscopic image analysis and settling column experiments, the settling velocity and size distribution of anaerobic granular sludge samples. The theoretical settling velocities were then determined using equivalent diameter. The settling velocity of granules from lab sized reactors and industries have been measured well using equivalent diameter. However, an increase in Reynolds number (Re) in the experiments can result in weaker agreement between measured and calculated values.

As shown, different aspects of sedimentation of particles in aquatic environments have been previously investigated in a number of numerical and laboratory studies. Despite this, there is no literature describing the simulation of non-circular particles using high-resolution numerical simulation with an equivalent diameter. This paper uses a 3D computer simulation to determine whether it is possible to replace non-circular particles with round equivalents. The derivation of the CFD and DEM equations and the coupling process is explained in the next section. Analysis of the simulation results and comparison with the lab measured data is provided in the Results section, and some concluding results complete this study.

2 Methodology

The methodology of this study is based on a numerical model that couples Computational Fluid Dynamics (CFD) and the Discrete Element Method (DEM).

The model used CFD to calculate hydrodynamics, a fixed Eulerian grid. As a result of numerical methods used in CFD, it is possible to predict velocities, pressures, and shear stresses with respect to flow fields through numerical approaches that can be used to forecast the flow fields. Having a simulation's prediction capability validated and verified is one of the most crucial steps in boosting trust in the simulation's ability to predict. The purpose of evaluating the accuracy of the numerical solutions to the governing equations is to determine whether the solutions are precise or not. The accuracy of the numerical solutions is determined by analyzing their mesh design and sensitivity, spatial and temporal discretization accuracy, as well as convergence criteria. As the name suggests, validation is the process by which one checks that the simulation accurately represents the world that is being simulated. There is a need for detailed information regarding key areas that are difficult to quantify or visualize, and it is important to anticipate factors that may be difficult to quantify, such as wall shear stress and particle residence time, which are likely to impact fluid flow in the future in ways that will be difficult to quantify [17].

The DEM is based on the Lagrangian method, where all particles are tracked by explicitly solving their trajectories [18]. As a result of its ability to resolve the granular medium at the particle size, the DEM is a valuable model. With the help of this method, it is possible to create realistic force chains for contacts, and thus one can observe phenomena such as particle segregation through percolation, which is influenced by the relative particle velocity and particle geometry. Hence, it becomes possible to capture a wide range of phenomena, including wave propagation inside granular material, equilibrium states, rapid and slow movements, denser and diluted particle regimes, and also wave propagation within granular material, as a result of this [19].

A promising method, the combined CFD-DEM technique, has been used to study granular-fluid systems. CFD-DEM is of interest because it has a superior computational efficiency over the Direct Numerical Simulation Discrete Element Method (DNS-DEM) and Lattice-Boltzmann Discrete Element Method (LB-DEM), in addition to its ability to capture particle physics more accurately than a two-fluid-model (TFM) [20]. Using CFD-DEM models, many phenomena can be simulated, including drag models, particle–particle interaction models, particle–wall interactions, simulations of biofilm growth and decay, saltation, rolling and sliding, particle aggregation, and simulation of erosion and sedimentation morphological changes that occur during erosion and sedimentation. However, the downside of this method is that it has several limitations, including a limited number of particles and due to a high computational cost, which makes it unsuitable for field-scale simulations because of these limitations. Recently, Kloss et al. [19] introduced a CFD-DEM coupling method known as CFDEM, 2011. The method utilizes the Open-Source software packages OpenFOAM® (OpenCFD Ltd., 2009) and LIGGGHTS (LIGGGHTS, 2011).

2.1 CFD Model

The Navier–Stokes equations for the incompressible fluid domain are integrated with boundary and initial conditions. This will assure the conservation of momentum and mass. These equations hold on the whole domain $\Omega = \Omega_F + \Omega_P$ (see Fig. 1). The governing equations include the momentum equation [21]:

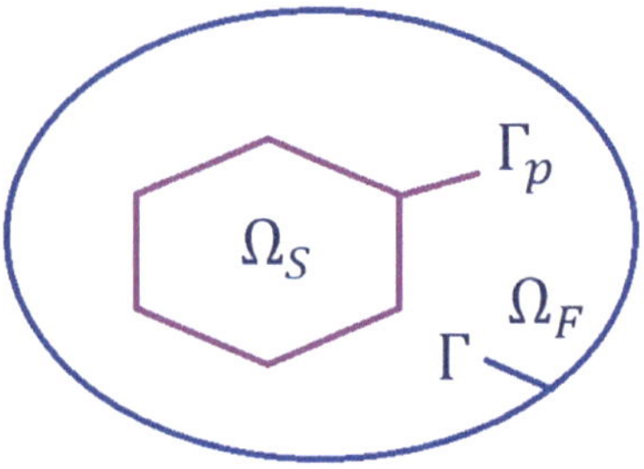

Fig. 1 Sketch of particle domain and fluid domain [19]

$$\rho_f \frac{\partial \boldsymbol{u}_f}{\partial t} + \rho_f (\boldsymbol{u}_f.\nabla)\boldsymbol{u}_f = -\nabla \mathrm{p} + \mu \nabla^2 \boldsymbol{u}_f \text{ in } \Omega_F \tag{1}$$

the continuity equation,

$$\nabla \cdot \boldsymbol{u}_f = 0 \text{ in } \Omega_F \tag{2}$$

with boundary conditions.

$$\boldsymbol{u}_f = \boldsymbol{u}_\Gamma \text{ on } \Gamma \tag{3}$$

and interface conditions on the fluid–solid boundary.

$$\boldsymbol{u}_f = \boldsymbol{u}_\mathrm{p} \text{ and } \sigma \cdot \hat{n} = \mathrm{t}_{\Gamma_p} \text{ on } \Gamma_p \tag{4}$$

as well as the initial conditions given by.

$$\boldsymbol{u}_f(x, t = 0) = \boldsymbol{u}_0(x) \text{ in } \Omega_F \tag{5}$$

where ρ_f and $\boldsymbol{u}_f$ are the fluid density and velocity, $\hat{n}$ is the outward normal unit vector, Γ_p is the fluid domain, Ω_F is the computational domain, σ is tension inside the fluid, $\boldsymbol{u}_\mathrm{p}$ is the velocity of particle, p is the dynamic pressure, μ is the fluid viscosity and t_{Γ_p} is the traction vector acting from the fluid on the solid body surface.

In order to solve the system of equations numerically using numerical methods, it is necessary to change the analytical form of the equations to the spatio-temporally discretized form given by the following equations [19]:

$$\rho_f \frac{\widehat{\boldsymbol{u}_f} - \boldsymbol{u}_f^{(n-1)}}{\Delta t} + \rho_f \left(\boldsymbol{u}_f^{(n-1)}.\nabla\right)\boldsymbol{u}_f^{(n-1)} = -\nabla \tilde{p} + \mu \nabla^2 \boldsymbol{u}_f^{(n-1)} \tag{6}$$

where, Δt is time-step size and $\tilde{p}$ means an estimated pressure value that will be corrected using the Pressure-Implicit with Splitting of Operators (PISO) algorithm [22] afterward, $\widehat{\boldsymbol{u}_f}$ is the interim solution, $\boldsymbol{u}_f^{(n-1)}$ is the previous time step velocity solution. To correct $\tilde{p}$, a Poisson equation can be used for the pressure. Correcting the velocity in the particle domain to the velocity taken from the DEM data or an input file is the next step. The velocity correction is equal to an additional force term in the Navier–Stokes equations. After convergence of the iterations, the value of the velocity at the present step can be calculated as $\boldsymbol{u}_f^{(n)} = \widehat{\boldsymbol{u}_f}$.

Technically, the suggested correction means the addition of a force term f in the Navier–Stokes equations, which meets the following criteria:

$$\rho_f \frac{\widetilde{\boldsymbol{u}_f} - \widehat{\boldsymbol{u}_f}}{\Delta t} = f \tag{7}$$

$$\boldsymbol{u}_f^{(n)} = \widetilde{\boldsymbol{u}_f} - \nabla\varphi \tag{8}$$

where φ is a scalar field, $\widehat{\boldsymbol{u}_f}$ is the interim velocity and $\widetilde{\boldsymbol{u}_f}$ is the current velocity. Note that the role of $\nabla\varphi$ is to fix the problem of violation of the divergence-free condition for the velocity over the whole domain.

2.2 *Implementation of Discrete Element Method (DEM)*

DEM is a strong method regarding its capacity to elucidate the granular medium, especially at the scale of particles. Thereby, it is capable of calculating accurate contact force along with considering the further situations caused by particle geometry and integrated relative particle motion. Thus, several cases can be captured including particle segregation by percolation, dense description and particulate regime dilution, a wide range of flow ranging from rapid to slow, as well as either wave propagation or equilibrium states within the granular substance. Additional details on the contact physics as well as implementation matters are accessible in the following literature [23, 24, 20].

2.2.1 DEM Model

DEM is considered to be a Lagrangian model since it tracks all the particles in the calculation domain using explicit solving of their trajectories. The particle force balance with index *i* reads.

$$m_i\ddot{\boldsymbol{x}}_i = \boldsymbol{F}_{i,n} + \boldsymbol{F}_{i,t} + \boldsymbol{F}_{i,f} + \boldsymbol{F}_{i,b} \tag{9}$$

$$I_i \frac{d\boldsymbol{\omega}_i}{\mathrm{d}t} = \boldsymbol{r}_{i,c} \times \boldsymbol{F}_{i,t} + \boldsymbol{T}_{i,r} \tag{10}$$

where I_i, $\boldsymbol{\omega}_i$, m_i and $\ddot{\boldsymbol{x}}_i$ are moment of inertia, rotation speed, mass and linear acceleration respectively for the ith particle. $\boldsymbol{F}_{i,n}$ is the normal particle–particle contact force, $\boldsymbol{F}_{i,t}$ is the tangential particle–particle contact force and $\boldsymbol{r}_{i,c}$ is the normal vector to $\boldsymbol{F}_{i,t}$ coinciding the center of particle. $\boldsymbol{F}_{i,f}$ is the force that the nearby fluid phase might apply on the particles, but it is not discussed here further. Other body forces such as electrostatic, magnetic, or gravitational forces are summarized as $\boldsymbol{F}_{i,b}$. $\boldsymbol{T}_{i,r}$ is an extra torque on the particle that can be used to simulate non-sphericity through a 'rolling friction' model (e.g., LIGGGHTS, 2011). Every particle is represented by a sphere mathematically, another geometrically distinct volume or a mixture of them. The related momentum balances are fundamental to both the translational and angular accelerations of each sphere.

2.2.2 Particle to Particle Contact Model

A linear spring-dashpot model is the most uncomplicated example, shown in Fig. 2. In terms of this model, the normal force is assumed by:

$$\boldsymbol{F}_n = -k_n \boldsymbol{\delta}_n + c_n \Delta \boldsymbol{u}_n \tag{11}$$

the tangential contact force can be calculated using:

$$\boldsymbol{F}_t = \min\left\{ \left| k_t \int_{t_{c,0}}^{t} \Delta \boldsymbol{u}_t \mathrm{d}t + c_t \Delta \boldsymbol{u}_t \right|, \mu \boldsymbol{F}_n \right\} \tag{12}$$

$\Delta \boldsymbol{u}_t$ is the particle's relative tangential velocity in contact. k_t, k_n, and c_t, c_n are the tangential and normal spring and damping coefficients, respectively, and $\boldsymbol{\delta}_n$ is the normal overlap. This formula can be similarly used for particle–wall contact. The tangential overlap is shortened to meet the Coulomb friction criterion with μ being the friction coefficient Kloss et al. [19].

The integral part of the Eq. 12 signifies an additive spring storing the energy of the relative tangential motion, showing the elastic tangential deformation of each particle's surface. This happens from the time the particles collide. The other part of Eq. 12, the dashpot, considers the energy indulgence of the tangential contact. The

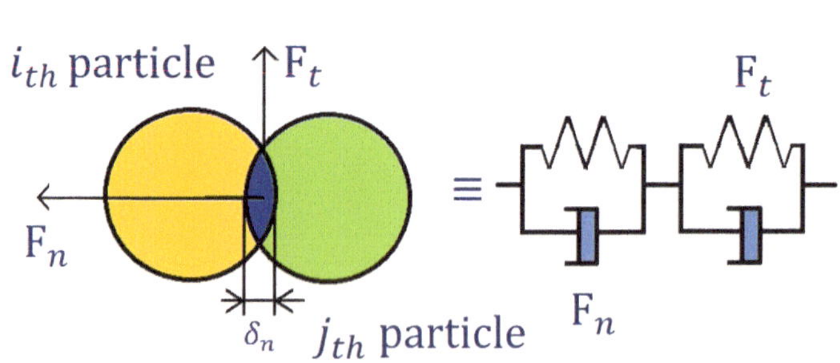

Fig. 2 Simple spring-dashpot model adapted from Yeom et al. [25]

Coulomb frictional limit restricts the amount of the tangential force, wherever the particles commence to slide over one another.

Various kinds of contact models or contact laws can be applied by altering k_t, k_n, c_t, and c_n as an overlap function (linear and non-linear).

2.2.3 Contact Detection Algorithm

Calculating the contact points between particles interacting at each time step is a time-consuming and critical component of the discrete element method. Therefore, it is necessary to choose a suitable contact detection algorithm. Developed by Verlet [26], the neighbor list was the first of its kind to be developed. The objective of this project is to develop a database of potential contacts on a regular basis and provide them with a list of those contacts. Afterwards, the list of particles that have made contact with each other is checked for each time-step, and evaluations can be drawn based on the actual contacts (such as the calculation of particle–particle forces) that have been made. By utilizing this idea, we are able to create a pair of particles that we can presume to be excluding from contact because they are located too far apart to be in contact with each other. The neighbor list contains a pair of (spherical) particles if:

$$\|\boldsymbol{x}_i - \boldsymbol{x}_j\| < r_i + r_j + s \tag{13}$$

where i and j denotes the ith and jth particle, r_i and r_j represent the radios of ith and jth particle and s represents the so-called Verlet or skin parameter. This parameter can be chosen at the discretion of the user, as long as certain constraints are met [26]. The validity period of the list is therefore determined by the validity period of the list. The list is valid for N_{Verlet} time-steps meeting the following conditions given a constant time-step Δt and a maximum particle velocity of magnitude $V_{\max}$:

$$N_{Verlet} = \frac{s}{2V_{\max}\Delta t} \tag{14}$$

For the next N_{Verlet} time-steps, the purpose of this list is to detect collisions before creating a new list, and it is used to check for collisions before creating a new one. In general, a neighbor list is created using a binning method, such as the link-cell method through grid-decomposition, in order to identify which particle pairs are qualified to be evaluated under Eq. 12 by using a binning method. The Verlet parameter s and the spacing of this binning are set for the algorithm's run-time. It is clear that the process becomes meaningless for both of the extreme values of s, zero and infinity. If $s = 0$, then no pairs will be included in the list. If $s = \infty$, then all pairs in the system would be included, declaring the algorithm useless. Since this run-time optimization is not a closed solution as it depends on specific hardware characteristics, such as the CPU cache size, as well as other factors, such as particle size distribution, particle shape, contact force implementation, flow regime, and so on, there is no closed solution to this run-time optimization.

2.3 Unresolved CFD-DEM Approach

Using a CFD-DEM model, each particle is taken into account separately, and a velocity and position for each particle are determined by integrating the forces acting on them over a period of time in order to estimate their velocity and position for each particle independently. However, despite this, a significant part of the process can be attributed to the size of the particles in comparison to the size of the grid. There are two main approaches to CFD-DEM: resolved and unresolved. When particle sizes are less than the size of the computational grid (cell), the unresolved technique is appropriate, and it should be noted that the particles are considered to partially fill the cell when their sizes are less than the computing grid. It would be appropriate to use the resolved approach instead when the particle size is very large and it will be necessary to cover more than one cell in the computational grid due to the particles covering more than one cell [19].

In the simulations performed in this study, the smallest ratio of grid size to particle was 63 and the biggest particle diameter was assumed to be d = 3.79 mm (volume = 11.28 mm^3). Based on the number of particles and the geometry, the highest volume fraction is 1/63 (0.016). To prevent any undesirable forces acting on particles or any unwanted interactions, we initially put each particle at enough distance to the others. The details can be found in *Geometry* section. This proves that unresolved CFD-DEM should be applied to the simulation. This approach deals with those cases containing particles whose sizes are smaller than the computational grid; hence, the cells are expected not to entirely brim over with a particle.

The volume-averaged Navier–Stokes equations can be used for incompressible-fluid phase motion in the sight of a secondary particulate phase, which can be written as:

$$\frac{\partial \alpha_f}{\partial t} + \nabla.(\alpha_f \boldsymbol{u}_f) = 0 \tag{15}$$

$$\frac{\partial (\alpha_f \boldsymbol{u}_f)}{\partial t} + \nabla.(\alpha_f \boldsymbol{u}_f \boldsymbol{u}_f) = -\alpha_f \nabla.\frac{p}{\rho_f} - \boldsymbol{R}_{pf} + \nabla.\boldsymbol{\tau} \tag{16}$$

where α_f is the fluid-occupied volume portion, ρ_f is fluid density, u_f is fluid velocity, and $\boldsymbol{\tau} = \nu_f \nabla \boldsymbol{u}_f$ is the fluid phase stress tensor. $\boldsymbol{R}_{pf}$ characterizes the momentum exchange via the particulate phase, calculated for every cell that is collected using the drag forces based on the particle. $\boldsymbol{R}_{pf}$ is calculated using:

$$\boldsymbol{R}_{pf} = \boldsymbol{K}_{pf}(\boldsymbol{u}_f - \boldsymbol{u}_p) \tag{17}$$

$$K_{pf} = \frac{\sum_i F_d}{V_{cell}.\left|u_f - u_p\right|} \quad (18)$$

where K_{pf} is momentum exchange coefficient, F_d is the force and u_p is averaged particle velocity.

In recent years, many more drag correlations have been presented for the K_{pf} calculation. [19] suggested the below formulations:

$$C_D = \frac{24}{\alpha_f Re_p}\left(1 + 0.15(\alpha_f Re_p)^{0.687}\right) \quad (19)$$

$$Re_p = \frac{\rho_f \left|u_f - u_p\right|}{\mu} d_p \quad (20)$$

where Re_p is the particle Reynolds number and d_p is the particle diameter. Now depending on α_f, the momentum exchange coefficient is calculated by:

$$K_{pf} = \frac{3}{4} C_D \frac{\alpha_f (1 - \alpha_f)\left|u_f - u_p\right|}{d_p} \alpha_f^{-2.65} \text{ if } \alpha_f > 0.8 \quad (21)$$

$$K_{pf} = 150\frac{(1 - \alpha_f)^2 \mu}{\alpha_f d_p^2} + 1.75\frac{(1 - \alpha_f)\left|u_f - u_p\right|}{d_p} \text{if } \alpha_f \le 0.8 \quad (22)$$

To solve the mentioned equations, a Pressure-Implicit with Splitting of Operators (PISO) pressure velocity coupling is applied [22]. As pressure and velocity are used as dependent variables in this approach, it can be used with equations of motion involving pressure and velocity as well. One of the key components of this technique is the division of the solution process into a number of steps, which is the key to its success. Because of the separated sets of equations that can be solved by standard methods, it is possible to decouple the operation on pressure from the operation on velocity at each step, thus decoupling the operation on velocity from the operation on pressure at each step.

The coupling procedure is as follows:

- The DEM solver determines the locations and velocities of the particles.
- The determined particle locations and velocities are transferred to the CFD solver.
- The corresponding cell in the CFD mesh is identified for each particle.
- The mean particle velocity and the particle volume fraction are calculated for each cell.
- The fluid forces acting on each particle are estimated based on the particle volume fraction.
- By ensemble averaging all of the particles in a CFD cell, particle–fluid momentum exchange terms are generated from particle-based forces.
- The DEM solver receives the predicted fluid forces acting on each particle and uses them in the next time step.

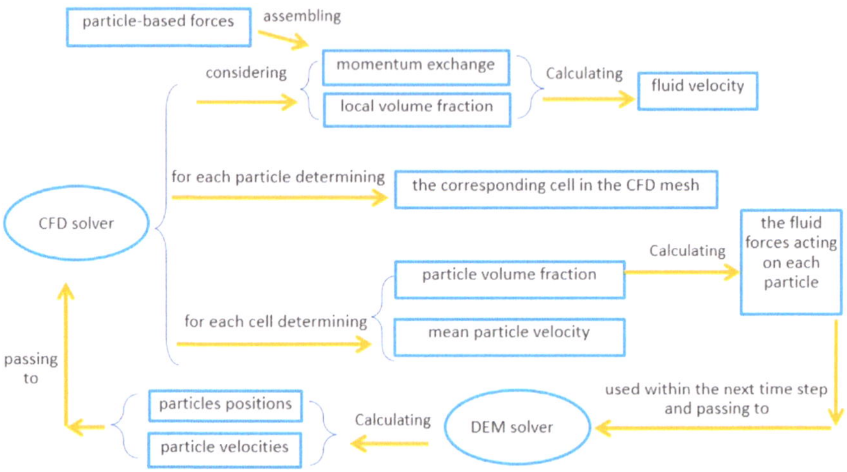

Fig. 3 Particle Fluid coupling procedure

- The fluid velocity is determined by the CFD solver while accounting for the local volume fraction and momentum exchange.
- Other equations might be added and evaluated.
- The procedure is repeated.

The coupling procedure consists of several steps as shown in Fig. 3.

Typically, DEM-time steps have to be smaller than CFD time steps due to particle collision dynamics as well as the necessities for maximum particle overlap aimed at particle collision modelling, especially when using the soft-sphere method. Regarding a flexible coupling scheme in hand, the CFD and DEM time-steps should be determined independently of each other. Moreover, it is possible to select a coupling interval in order to adjust to the quantity of physical coupling among the phases.

It is required to consider the volumes of the particles, a so-called Lagrangian property, from the DEM side to calculate the volume fraction field, which is set on the CFD side with a fixed Eulerian grid. These results from the aforementioned calculations of the particle volume fraction as well as a mean particle velocity which are essential to understanding the coupling.

As soon as the particle volume fraction calculation is done, evaluation of each particle's contribution to particle–fluid momentum exchange is possible. The tool to establish this is a drag force based on the granular volume fraction [19]. The CFDEM software was used to solve the above-mentioned equations.

2.4 *Simulation Conditions*

2.4.1 Geometry

Based on the obtained experimental data from unbonded sedimentation or floating velocities of different types of plastic pellets in water performed by Wang et al. [27] in a glass cylindrical chamber 50 cm in height and 7.90 cm in diameter, the average velocity of at least 200 pellets of each type of plastics has been obtained. As part of the investigation, three different stages of pellet were examined. In two stages of the experiment, the pellets are sat in artificial seawater and deionized water, and then their settling/floating velocities are measured. At the third stage of the process, it is important to take into account the settling/floating velocities of coated pellets. Due to these different conditions, there is a range of the terminal velocity for each particle instead of a single value. Currently, we consider a rectangular container as shown in Fig. 4. The computational domain for CFD-DEM had the initial dimensions of 15 cm × 50 cm × 15 cm, with the domain being divided into 16 × 52 × 16 grids. For each particle size in each simulation, there were 20 particles released in the particle release box (the blue one in the container) with no velocity at t = 0. The size of the release box was 8 cm × 4 cm × 8 cm. Note that for the sinking particles, the release box center was located 8 cm under the container ceiling, while for floating debris, the release box center was located 6 cm above the container base. In order not to interact with the walls, the base centers of the release box and the container were collinear; allowing each particle to be far away from the walls, at least 3.5 cm in the x and z directions.

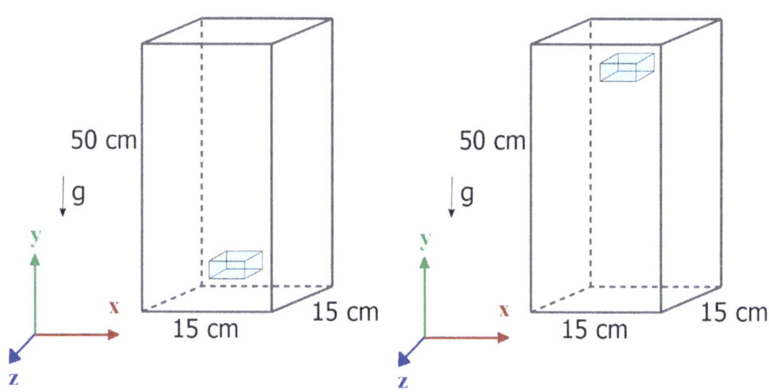

Fig. 4 Container and particle release box (blue box); for lighter than water (left) and for denser than water (right)

2.4.2 Grid Convergence

The study of grid convergence is a very important part of the study in order to confirm that numerical solutions are accurate approximations of the governing partial differential equations representing the problem under investigation.

The analysis should demonstrate that the spatial as well as the temporal discretization errors will asymptotically approach zero when the grid is gradually refined by increasing the number of cells and decreasing the cell sizes and time steps as the grid is successively refined. The computer round-off errors, which are minimal enough to be disregarded, are not included in the discretization error [28].

In this study, we use the relative difference between subsequent solutions as follows:

$$\epsilon = \frac{V^{j+1}_{avg} - V^{j}_{avg}}{V^{j}_{avg}} \tag{23}$$

$$V^{j}_{avg} = \frac{\sum_{i=1}^{20} V_{p_i}}{20} \tag{24}$$

where V_{p_i} is the terminal velocity of *ith* particle, ϵ is the relative difference between subsequent solutions, number 20 is the quantity of particles in each simulation and V^{j}_{avg} is the average terminal velocities according to the *j*th refinement step. The velocities in the experimental work are not exact and represent a range of values for the velocities in the experiment. Therefore, the average velocity of the particles is considered in the calculations. Table 1 illustrates the relative difference between subsequent solutions regarding our grid sizes of simulations:

As the relative difference between subsequent solutions for $j = 2$ is 10^{-6}, it points out that the average velocity difference between $j = 1$ and $j = 2$ is of the order of 10^{-6}. Therefore, we choose to use associated $j = 1$ grid size ($16 \times 52 \times 16$) for our simulations.

2.4.3 Coupling Properties

CFDEM® coupling provides an open source parallel coupled CFD-DEM framework combining the strengths of LIGGGHTS® DEM code and the Open Source CFD package OpenFOAM® [19]. In the "Coupling Properties" of CFDEM, the setup of the coupling routines of the CFD-DEM simulation are defined. It contains two parts: sub-models & settings and sub-model properties. In sub-models & settings, the routines are defined in Table 2 which are adapted from [29]. Sub-models and related properties of the model are presented in Table 2 in order to paint a generalized picture regarding simulation specifications.

The Model Type refers to the formulation of the equations to be solved. Model type A uses the local average method to directly derive the fluid governing equation on the basis of the point equation of motion of the fluid, and the solid phase governing equation on the basis of the equation of motion for the center of mass of a single

Table 1 Grid convergence based on the relative difference between subsequent solutions

j	Grid size	ϵ	Dimensions (cm)
0	8 × 26 × 8	10^{-1}	15 × 50 × 15
1	16 × 52 × 16	10^{-2}	
2	24 × 78 × 24	10^{-6}	

Table 2 Coupling properties

Model type	A	Mom couple models	Implicit cople
Coupling interval	50	Force models	gradPForce
Void fraction model	Divided		
Locate model	Engine		viscForce
Data exchange model	TwoWayMPI		Archimedes
Averaging model	Dense		KochHillDrag

particle. However, when the system of interest contains a significant number of closely spaced particles, the problem is too complex to allow for a direct solution. If it is stated under these conditions, it becomes clear that it has to be simplified in some way in order to be adequately characterized by a limited number of partial differential equations in order to be able to describe it adequately. Therefore, the derived equations can describe both the fluid motion and the particle motion as though they were interpenetrating continuums [30]. The Coupling Interval determines the time passing between two CFD-DEM data exchanges. The best method to solve the pure DEM problem is to first set the DEM timestep in accordance with the requirements of the pure DEM problem in order to find the best solution. Initially, the coupling interval can be short (short coupling interval) or it can be loose (long coupling interval) depending on the question. Secondly, the CFD timestep should be chosen. It must be equal to or smaller than the coupling time which is the product of DEM timestep and coupling interval. The divided void Fraction model is supposed to be used when a particle (or its representation) is in the size range of a CFD cell. The particle has a radius of R, and its volume is partitioned into 29 distinct, equal-sized parts that don't overlap. Then, each of these volumes is reproduced using its centroids. The first volume has a sphere-like shape, with the particle center coinciding with its center. This subsphere's radius r_1 can be found as follows:

$$\frac{V(r_1)}{V(R)} = \frac{r_1^3}{R^3} = \frac{1}{29} \rightarrow R^3\sqrt{1/29} \tag{25}$$

where V is the volume and R is the radius of particle. A spherical layer known as the rest volume must be separated into two layers of equal volume. It is simple to determine where the border between these two spherical layers is in the radial direction:

$$\frac{V(r_1)}{V(R)} = \frac{r_2^3}{R^3} = \frac{15}{29} \rightarrow R^3\sqrt{15/29} \tag{26}$$

where r_2 is the radius of the remaining volume. Each of these spherical layers is subsequently separated into 14 equal-volume components. Now for the first spherical layer we have:

$$r_{S,1} = \frac{\int rdV}{\int dV} = \frac{\int_{r_1}^{r_2} r^3 dr}{\int_{r_1}^{r_2} r^2 dr} = \frac{3}{4}\frac{r_2^4 - r_1^4}{r_2^3 - r_1^3} \approx 0.62761\text{R} \tag{27}$$

where $r_{S,1}$ is the position of the centroid point for each part in the first spherical layer. The same logic applies to the second spherical layer, keeping in mind that the particle radius is the exterior radius:

$$r_{S,2} = \frac{3}{4}\frac{R^4 - r_2^4}{R^3 - r_2^3} \approx 0.90853\text{R} \tag{28}$$

where $r_{S,2}$ is the position of the centroid point for each part in the secpnd spherical layer. The "weight" parameter can be defined and can modify the particle volume that is occupied in the CFD domain using:

$$V_p = \frac{4}{3}\pi r^3.w \tag{29}$$

where w is the weight parameter. Note that the weight parameter is dimensionless. The base class for models that look for the CFD cell and cell-ID corresponding to a point is called locate Model. It is typically used to determine which cell a particle is in. for the engine module allow different settings to be utilized to employ various algorithms.

The Data Exchange Model performs the data exchange between LIGGGHTS and within CFDEM coupling toolbox. The exchanged data at least consists of positions, radii, velocities and forces. The twoWayMPI model can exchange particle properties from DEM to CFD and from CFD to DEM.

The Averaging Model performs the Lagrangian to Eulerian mapping of data (e.g. particle velocities). This averaging model is used to calculate the average particle velocity inside a CFD cell. The "dense" model is supposed to be applied to cases where the granular regime is rather dense.

The Mom Couple Models is the base class for momentum exchange between DEM and CFD simulation. The implicitCouple model is the one providing an implicit momentum source term for the CFD solver. Forces may be implicitly connected to the fluid solver (i.e., the fluid velocity at the new time will be taken into account for the coupling force while solving the Navier–Stokes equations). Usually, drag forces are subject to this implicit coupling. Implicit coupling is more stable (particularly significant in dense flows). In the drag model implementation, explicit forces are applied to the flow solver in a way that is less stable.

The Force Models perform the calculation of forces (e.g. fluid-particle interaction forces) acting on each DEM particle and on each CFD cell. All force models selected are executed sequentially and the forces on the particles are superposed. The majority of force models can be employed with coarse graining. The gradPForce model is a model that calculates the particle-based pressure gradient force. The viscForce model calculates the particle based viscous force. The KochHillDrag model is a model that calculates the particle-based drag force following the correlation of Koch and Hill [31]. The Archimedes model is a model that calculates the Archimedes' volumetric lift force stemming from the density difference of fluid and particle.

3 Results

3.1 Parameters

As can be seen in Table 3, there are some common particle specifications as well as fluid parameters for each simulation. They have been defined in the DEM and CFD sections of the simulation, respectively. These parameters were constant during the simulation which are adapted from [29].

Note that to obtain the terminal velocity, which is a constant quantity according to its definition, each simulation has been run for 20 s of real time. The real run takes 45 min, which is a reasonable amount of CPU time. We take the measured velocity into account after the first 3 s to be sure that the velocity becomes constant. In order to validate the simulation with experimental data, the simulations are restricted to Table 3

Figure 5 shows the result of one of the simulations. The result has been visualized by ParaView software. The simulation is for HDPE particle which is lighter than water. According to the description of geometry in Sect. 2.4.1, particles are placed at the bottom of the container. As the result showing $t = 5$ s, particles have been moving upward due to buoyancy force. This movement caused a minor fluid velocity. The velocity of fluid shows that our simulation considered particle–fluid interactions.Our simulation results are in good agreement (6 out of 8 simulations met

Table 3 Simulation parameters

Particle		Fluid	
Parameter	Value (unit)	Parameter	Value (unit)
Young module	5×106 (N/m^2)	Transport model	Newtonian
Restitution coefficient	0.3	Time step	5×10^{-4} (s)
Time step	10^{-5} (s)	Density	1022 (kg/m^3)
Poisson ratio	0.45	Viscosity	10^{-6} (m^2/s)
Friction coefficient	0.5	Regime	Laminar

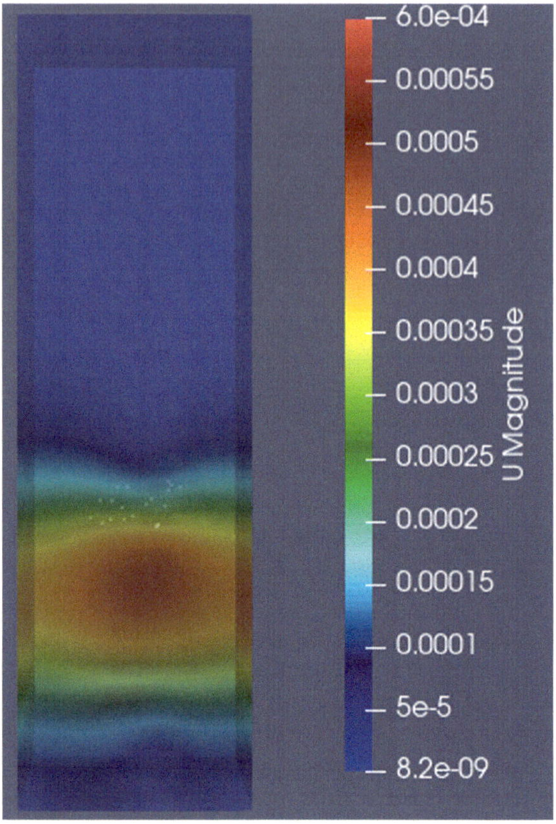

Fig. 5 The simulation at t = 5 s for high density polyethylene (HDPE) using ParaView software. "U magnitude" is the fluid velocity (m/s)

the experimental data) with experimental results for millimeter-size round particles. It can be seen in Figs. 6, 7, 8, 9, 10 and 11 that the terminal velocity (either for floating or sedimentation) obtained by our simulation is in the range measured by the experimental setup.

As can be observed, the terminal velocities of polypropylene (PP), polystyrene (PS), polyethylene terephthalate (PET), high density polyethylene (HDPE), and two types of silicone rubbers, one dense and the other light, were simulated in water. The particle shapes of the denser ones, PET, Silicone and PS are cube, hemisphere, and ellipsoid sphere respectively. HDPE, PP and Silicone shapes are ellipsoid-short cylinder, hemisphere, and ellipsoid sphere. Regarding average particle velocity, HDPE with a short cylinder shape has the least variation (deviation per average rate is 0.62%), while PS with an ellipsoid sphere shape has the most fluctuation (deviation per average rate is 2.3%). Therefore, it is not possible to estimate their terminal velocity only by their shape. Horizontal directions illustrate the number of particles in Figs. 6, 7, 8, 9, 10 and 11.

On the other hand, our simulation results do not fully conform to experimental results for two kinds of particles: Polyurethane (PUR) and Polyvinyl Chloride (PVC).

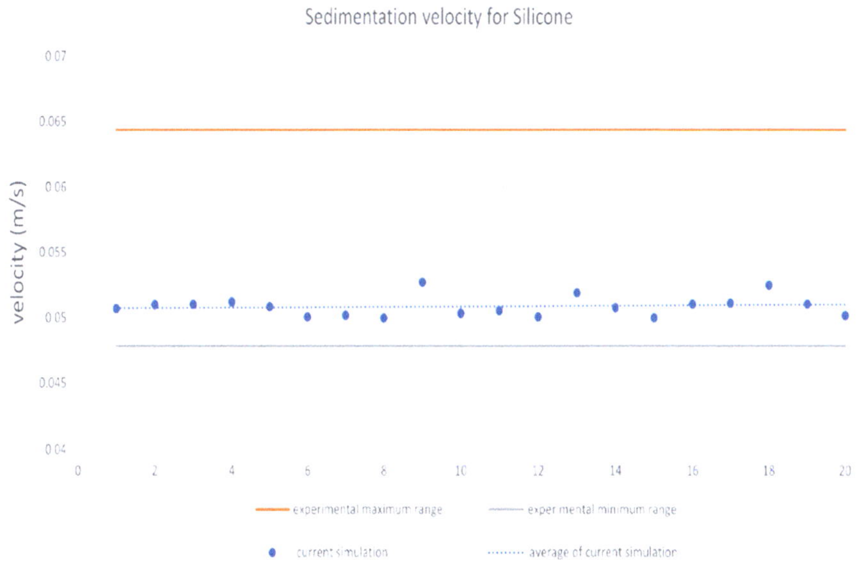

Fig. 6 Sedimentation velocity for Silicone ($D_e = 3.12$ mm and density $= 1139$ kg/m^3)

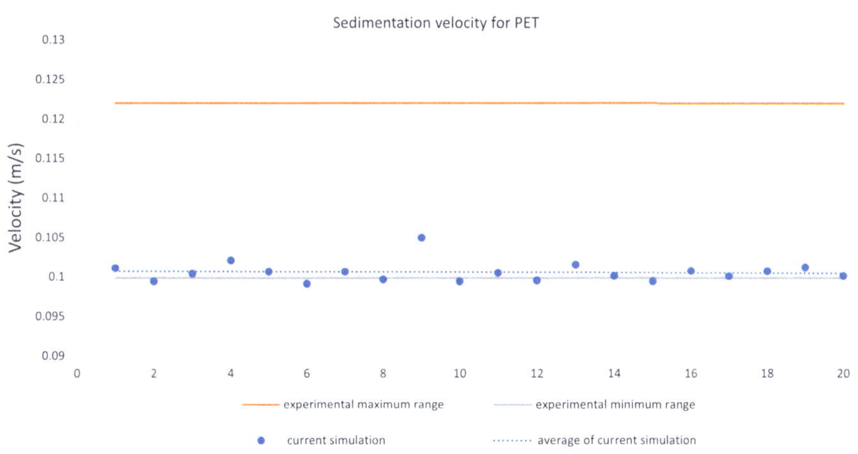

Fig. 7 Sedimentation velocity for PET ($D_e = 3.64$ mm and density $= 1353$ kg/m^3)

As can be seen in Figs. 12 and 13, the terminal velocity obtained by our simulation is lower than the minimum calculated range by the experimental setup.

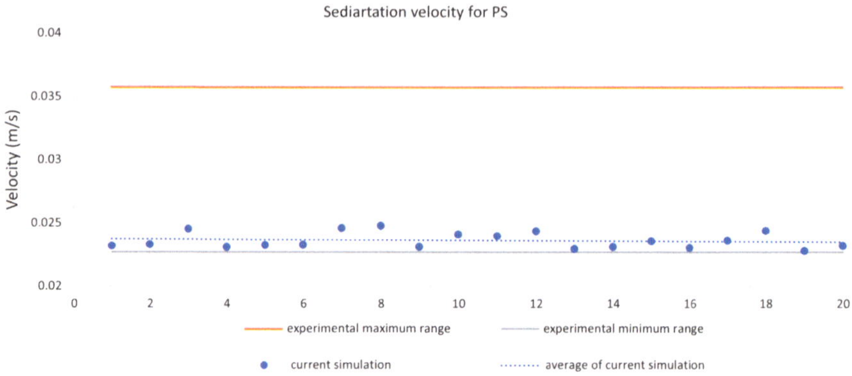

Fig. 8 Sedimentation velocity for PS ($D_e = 3.79$ mm and density = 1056 kg/m^3)

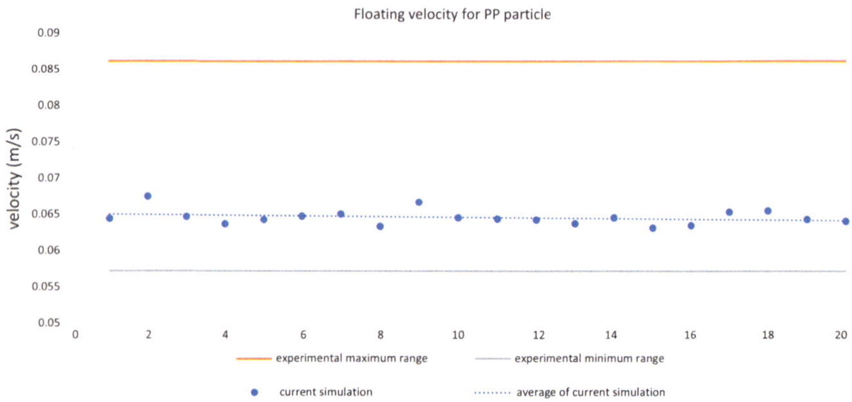

Fig. 9 Floating velocity for PP ($D_e = 3.70$ mm and density = 866 kg/m^3)

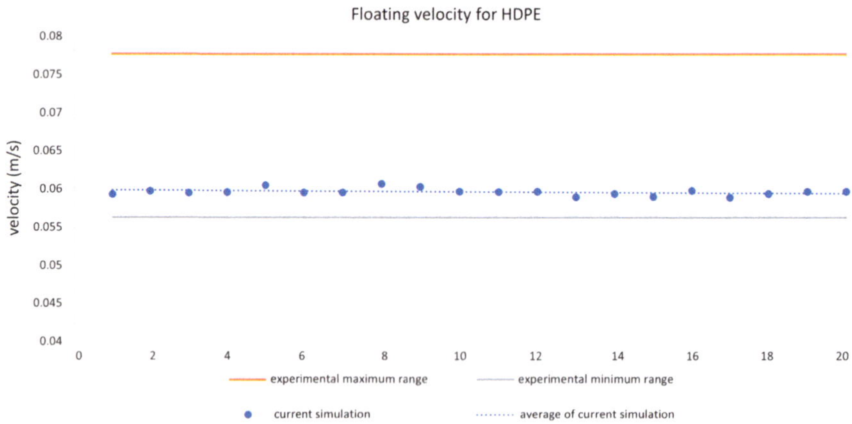

Fig. 10 Floating velocity for HDPE ($D_e = 3.01$ mm and density = 920 kg/m^3)

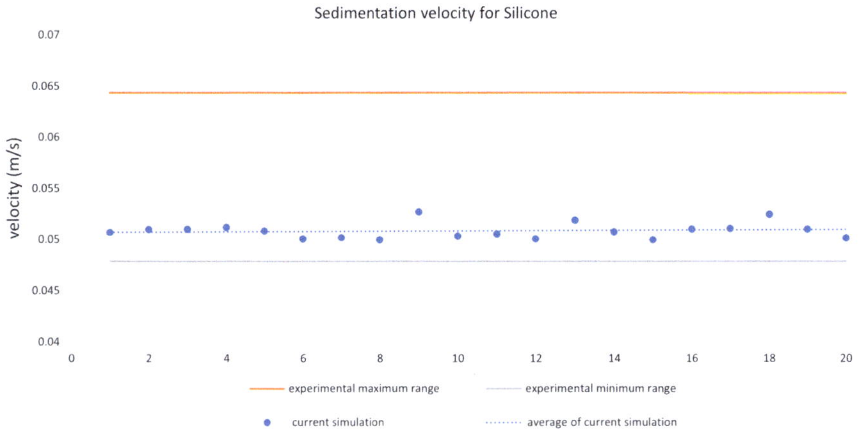

Fig. 11 Floating velocity for Silicone (D_e = 3.13 mm and density = 882 kg/m^3)

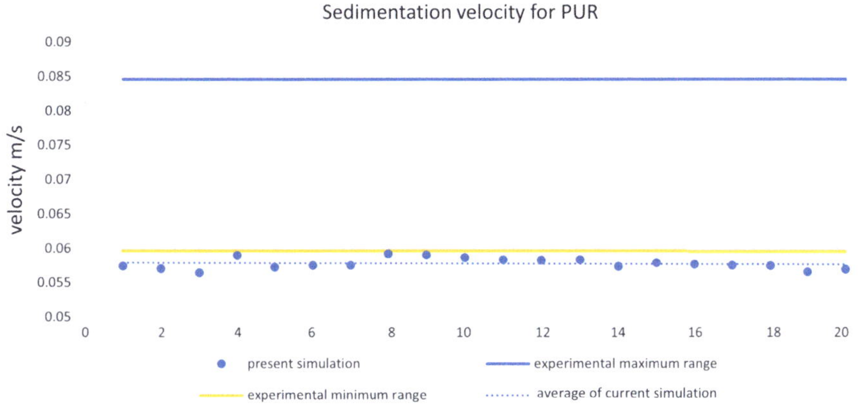

Fig. 12 Floating velocity for PUR (D_e = 2.85 mm and density = 1143 kg/m^3)

4 Discussion

In general, the velocity range for each particle is smaller than in the experimental setup. One of the causes for this shorter range would be the smaller number of particles in comparison with the experimental setup. In the experimental setup, there were 200 particles released in a cylindrical container whose volume is 2450 cm^3 (volume per particle rate is 12.25), while our simulation container is a cube with a volume of 11,250 cm^3 with only 20 particles (volume per particle rate is 562.5). We chose only 20 particles for each simulation and a relatively bigger grid size to particle ratio as well as more space for each particle just to make sure that there is no effective interaction among particles. This has guaranteed that particle–particle interactions cannot affect their terminal velocity. Also, it can be observed that the

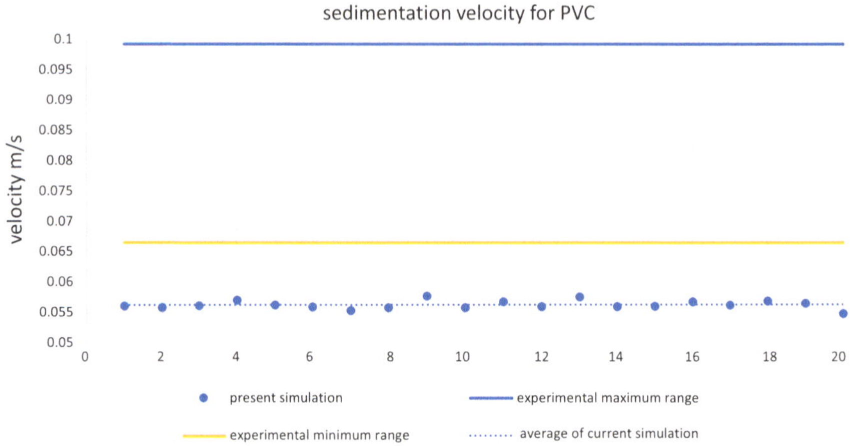

Fig. 13 Floating velocity for PVC (D_e = 2.69 mm and density = 1147 kg/m^3)

simulated terminal velocity averages are below the experimental terminal velocity average.

Besides, simulations of terminal velocities in our study for two particles did not agree with the experimental setup. These particles are Polyurethane (PUR) and Polyvinyl Chloride (PVC) whose densities and equivalent diameters are 1143, 1147 kg/m^3, 2.85 and 2.34 mm respectively. Their shapes are Olivary and Hemi-spherical Cylinder. The reason for this disagreement between our simulation and the lab setup could be that the presumption of equivalent diameters was inaccurate. In addition, another explanation could be that there might be some kind of interaction between the volume per particle rate and the experimental setup.

5 Conclusions

The present study models the movement of solid particles in liquid (sinking or floating depending on their densities) by means of Computational Fluid Dynamics-Discrete Element Method (CFD-DEM). In this work, we used this coupling algorithm in order to represent the movements in a quiescent container. Taking into account particle sizes ranging from 1 to 5 mm, a variety of particles were simulated using CFDEM software. Comparing our simulations to prior experiments, it was found that our modelling agrees with prior experiments in 75% of the cases for the range of particles mentioned. In light of the complexity of the problem, it seems this outcome would be a reasonable one. Additionally, it is important to note that one of the unique aspects of this study is that a validated numerical setup was developed in order to be able to examine the terminal velocities of pellets. It is a setup that is both timely from the point of data collection and cost-effective from the point of view of utilizing various tools for measuring. It may be possible to continue particle research using the framework presented here, which has already been validated in a laboratory setting. Studies can be directed in a wide variety of ways in the future, depending on the focus of the study. There is potential for future developments in the field of non-spherical particle behavior, fluids, and the effect of temperature on the terminal velocity of particles.

References

1. G. Bagheri, C. Bonadonna, Aerodynamics of volcanic particles: characterization of size, shape, and settling velocity, in *Volcanic* Ash (Elsevier, 2016), pp. 39–52
2. J. Saxby, A. Rust, F. Beckett, K. Cashman, H. Rodger, Estimating the 3D shape of volcanic ash to better understand sedimentation processes and improve atmospheric dispersion modelling. Earth Planet. Sci. Lett. **534**, 116075 (2020)
3. M. Fan, D. Su, L. Yang, Development of a benchmark for drag correlations of nonspherical particles based on settling experiments of super-ellipsoidal particles. Powder Technol. **409**, 117811 (2022)
4. S. Yokojima, R. Takashima, H. Asada, T. Miyahara, Impacts of particle shape on sedimentation of particles. Eur J Mech B/Fluids **89**, 323–331 (2021)
5. C. Zhou, J. Su, H. Chen, Z. Shi, Terminal velocity and drag coefficient models for disc-shaped particles based on the imaging experiment. Powder Technol. **398**, 117062 (2022)
6. J.F. Chen, Y. Luo, J.H. Xu, Q.M. Chen, J. Guo, Visualization study on sedimentation of micron iron oxide particles. J. Colloid Interface Sci. **301**(2), 549–553 (2006)
7. S. Karimnejad, A.A. Delouei, M. Nazari, M.M. Shahmardan, A.A. Mohamad, Sedimentation of elliptical particles using Immersed Boundary-Lattice Boltzmann Method: a complementary repulsive force model. J. Mol. Liq. **262**, 180–193 (2018)
8. Y. Zhou, H. Ni, Z. Shen, M. Zhao, Study on particle settling in supercritical carbon dioxide drilling and fracturing. J. Petrol. Sci. Eng. **190**, 107061 (2020)
9. B. Hwang, H.S. Park, W.H. Jung, M. Lee, M.H. Kim, Numerical validation and investigation for the sedimentation of solid particles in liquid pool using the CFD-DEM coupling algorithm. Nucl. Eng. Des. **355**, 110364 (2019)
10. Z. Xie, Y. Shen, K. Takabatake, A. Yamaguchi, M. Sakai, Coarse-grained DEM study of solids sedimentation in water. Powder Technol. **361**, 21–32 (2020)
11. I. Isachenko, Catching the variety: obtaining the distribution of terminal velocities of microplastics particles in a stagnant fluid by a stochastic simulation. Mar. Pollut. Bull. **159**, 111464 (2020)
12. Q. Huang, Y. Ma, Y. Zhang, G. Pan, Numerical investigation of movement patterns of particles falling in a viscous fluid. Mech. Res. Commun. **119**, 103814 (2022)
13. H. Kalman, Effect of particle shape on void fraction. Powder Technol. **407**, 117665 (2022)
14. I. Fonceca, D. Maza, R.C. Hidalgo, Modeling particle-fluid interaction in a coupled CFD-DEM framework, in *EPJ Web of Conferences*, vol. 249 (EDP Sciences, 2021), p. 09004
15. A. Hazzab, A. Terfous, A. Ghenaim, Measurement and modeling of the settling velocity of isometric particles. Powder Technol. **184**(1), 105–113 (2008)
16. F.A. Tassew, W.H. Bergland, C. Dinamarca, R. Bakke, Settling velocity and size distribution measurement of anaerobic granular sludge using microscopic image analysis. J. Microbiol. Methods **159**, 81–90 (2019)
17. R.A. Malinauskas, P. Hariharan, S.W. Day, L.H. Herbertson, M. Buesen, U. Steinseifer, B.A. Craven, FDA benchmark medical device flow models for CFD validation. ASAIO J. **63**(2), 150–160 (2017)

18. E. Kosari, K. Vafai, Thermal stimulation of targeted neural circuits via remotely controlled nano-transducers: a therapy for neurodegenerative disorders, in *Advances in Heat Transfer*, vol. 52. (Elsevier, 2020), pp. 543–581
19. C. Kloss, C. Goniva, A. Hager, S. Amberger, S. Pirker, Models, algorithms and validation for opensource DEM and CFD–DEM. Prog. Comput. Fluid Dyn. Int. J. **12**(2–3), 140–152 (2012)
20. H.P. Zhu, Z.Y. Zhou, R.Y. Yang, A.B. Yu, Discrete particle simulation of particulate systems: theoretical developments. Chem. Eng. Sci. **62**(13), 3378–3396 (2007)
21. A.A. Shirgaonkar, M.A. MacIver, N.A. Patankar, A new mathematical formulation and fast algorithm for fully resolved simulation of self-propulsion. J. Comput. Phys. **228**(7), 2366–2390 (2009)
22. R.I. Issa, Solution of the implicitly discretised fluid flow equations by operator-splitting. J. Comput. Phys. **62**(1), 40–65 (1986)
23. A. Di Renzo, F.P. Di Maio, Comparison of contact-force models for the simulation of collisions in DEM-based granular flow codes. Chem. Eng. Sci. **59**(3), 525–541 (2004)
24. T. Pöschel, T. Schwager, *Computational Granular Dynamics: Models and Algorithms* (Springer Science & Business Media, 2005)
25. S.B. Yeom, E.S. Ha, M.S. Kim, S.H. Jeong, S.J. Hwang, D.H. Choi, Application of the discrete element method for manufacturing process simulation in the pharmaceutical industry. Pharmaceutics **11**(8), 414 (2019)
26. L. Verlet, Computerexperiments on classical fluids. I. Thermodynamical properties of Lennard-Jones molecules. Phys. Rev. **159**(1), 98 (1967)
27. Z. Wang, A. Pilechi, M. Cheung, P. Ariya, On the importance of dynamic and physicochemical processes on sedimentation: an integrated experimental-modeling study of microplastics in water. Environ. Sci. Technol. J. (submitted) (2022)
28. L. Kwaśniewski, Application of grid convergence index in FE computation. Bull. Pol. Acad. Sci. Techn. Sci. (1) (2013)
29. Z.Y. Zhou, S.B. Kuang, K.W. Chu, A. Yu, Discrete particle simulation of particle–fluid flow: model formulations and their applicability. J. Fluid Mech. **661**, 482–510 (2010)
30. T.B. Anderson, R. Jackson, Fluid mechanical description of fluidized beds. Equations of motion. Ind. Eng. Chem. Fundam. **6**(4), 527–539 (1967)
31. D.L. Koch, R.J. Hill, Intertial effects in suspension and porous-media flows. Annu. Rev. Fluid Mech. **33**, 619 (2001)

A Central Finite Volume Surface Gradient Method for the Two-Layer Shallow Water Equations with Rigid Lid

Rony Touma and Ralph Lteif

Abstract We develop here a new formulation of the two-layer shallow water equations with rigid-lid on variable-bottom topography. This formulation is presented as a conservative system where the function describing the geometry of the waterbed or bottom topography is embedded within the divergence of the flux term. The resulting system is then solved using an adaptation of the unstaggered central scheme (UCS) we previously developed. To satisfy the lake-at-rest physical constraint, we blend the surface gradient method (SGM) with our UCS scheme. The resulting UCS-SGM scheme is then validated and successfully implemented, representing an extension of the method previously presented for shallow water equations (one-layer flow) to address the complexities of two-layer rigid-lid flows while ensuring well-balanced characteristics. The performed numerical tests confirm the potential of the proposed method to handle two-layer flow problems on either smooth or non-smooth bottom topographies and also to handle steady- and unsteady-state flow problems.

1 Introduction

In this paper, we study the propagation of internal waves at the interface between two layers of fluids of different densities. Internal waves have an essential effect on ocean motion, equilibrium of marine ecosystems, and climatic mechanisms, see [30].

Over the years, many researchers have been interested in studying the mathematical aspect of the internal waves. In fact, the governing equations of the two-layer flow

R. Touma (✉)
Department of Computer Science and Mathematics, Lebanese American University, Byblos, Lebanon
e-mail: rony.touma@lau.edu.lb

R. Lteif
School of Arts and Sciences, Computer Science and Mathematics Department, Lebanese American University, Beirut, Lebanon
e-mail: ralph.lteif@lau.edu.lb

D. Zeidan et al. (eds.), *Computational Fluid Dynamics: Novel Numerical and Computational Approaches*, Infosys Science Foundation Series,
https://doi.org/10.1007/978-981-97-8152-2_2

so-called "*full Euler system* " are simple to obtain, but their mathematical and numerical study remain very difficult, see [7, 21, 23]. At this point, many researchers had interest in looking for simplified approximate asymptotic models that can be derived from the asymptotic expansions of existing non-local operators in the *full Euler system*, see [3, 7, 21, 23, 24]. To this end, the study is restricted to specific physical regimes thanks to smallness assumptions made on given dimensionless parameters. In this paper, we restrict ourselves to the widely known shallow water equations (SWE) scaling regime where the depth of the two layers of fluids is small compared with the internal wavelength. Many models have been rigorously studied in the SWE regime in [7, 14, 16]. To ease the reading of this paper, we do not try to derive here the "*full Euler system*" or the SWE from classical asymptotic analysis, the interested reader can see for instance [23]. Instead, we start from the commonly known two-layer SWE with a free surface [11, 45], and we derive the equations in the rigid-lid configuration.

Indeed, we confine our work to a commonly used approximation in mathematical modeling, namely, the rigid-lid assumption. This assumption is one of the principle assumptions made in the oceanographic community studying internal wave motion and is considered valid when the pycnocline is far from the water top surface [26, 40]. In fact, if the waves are deep enough relative to the wavelength of the disturbance, their interactions with the free surface are not significant and thus the top of the fluid can be viewed as an impenetrable, bounding surface. When analyzing the data of their experiments, Koop and Butler [33] assumed that the upper free surface can be considered as rigid. This assumption is valid only for small density contrast between the two fluids. This fact has been addressed in [31, 41] and established in more details in [20, 22, 45]. The flat rigid-lid assumption is frequently used as a significant simplification in the study of oceanographic two-layer flow and leads to a similar behavior to the free surface configuration when both fluids have comparable densities. Nonlinear models describing the propagation of internal waves delimited from above by a rigid lid can be found, for example, in [5, 7, 14, 37, 43, 44]. The validity of the rigid-lid approximation has been investigated in many works see, for instance, [16, 17, 19, 28, 35]. Using this assumption, the two-layer shallow water system consisting of two differential equations has a conservative structure suitable for numerical implementation. In fact, the numerical advantages of the rigid-lid formulation are considerable in the variable topography case. First, the model is conservative (i.e., no source terms) thus less numerical efforts are needed to maintain well balancing in order to preserve steady-state solutions, contrarily to the free surface configuration. Secondly, the system's eigenstructure can be explicitly obtained. This is not the case in the free surface configuration. A numerical scheme is considered to be well balanced [27] if it is able to preserve steady states at the discrete level. Using different mathematical techniques, many well-balanced numerical schemes for two layer with a free surface shallow water flows have been reported in literature, see [1, 11, 12, 34, 38, 39]. Well-balanced schemes for two-layer flows under the rigid-lid assumption have been less investigated, see [9, 36].

The robustness of a numerical scheme is measured by its ability to preserve steady-state solutions. Hence, in this paper, in order to solve numerically the two-layer rigid-lid flow problem, we propose to apply the surface gradient method (SGM), [52] within a central finite volume scheme. The surface gradient method is a general scheme for treating source terms based on an accurate reconstruction of the conservative variables at cell interfaces. Unfortunately, the rigid-lid SWE coupling the two variables (ζ, v) representing, respectively, the interface deformation and the shear mean velocity is not suitable to the application of the SGM.

In fact, the data reconstruction method based on conservative variables (ζ, v) does not provide accurate values on the computational control cells since the effect of bottom topography is not included. To this end, we give in this paper an equivalent formulation of the SWE as a system of three evolution equations on h_1, h_2, and v representing, respectively, the height of upper and lower layers and the shear mean velocity. This formulation is, in particular, suitable for the application of the surface gradient method (SGM). In fact, the piecewise linear reconstruction of the conservative variable $h_2(t, x) = d_2 + \zeta(t, x) - b(x)$ (where $b(x)$ represents the deviation of the bottom topography with respect to its rest state $z = -d_2$) provides accurate values on the cells since the interface level function $\zeta(t, x)$ is chosen as the basis and thus the effects of topography slope will be included.

In this paper, we apply a hybrid finite volume-surface gradient method for the numerical solution of two-layer rigid-lid systems. The finite volume method that we employ is based on our previously developed unstaggered central scheme (UCS) [46]; it evolves a piecewise linear numerical solution on a single grid, but uses a layer of staggered cells at an intermediate step to refrain from computing the solution of the Riemann problems arising at neighboring cells. Central schemes originally established in [42] mimic the staggered first-order method of Lax–Friedrichs by refraining from solving Riemann problems appearing at the cell interfaces. The UCS method is considered as an extension of the Nessyahu–Tadmor scheme [42] and is characterized by its simplicity and robustness. In order to avoid solving Riemann problems appearing at the cell interfaces, a numerical solution is computed on dual cells, then a backward projection step is applied to retrieve the solution values on the main grid. Furthermore, the UCS scheme applies a slope-limiting procedure for the forward and backward projection steps as well as for the predictor corrector step to ensure a non-oscillatory solution around shock waves. The UCS scheme was first developed for hyperbolic systems of conservation laws and then extended to the case of non-homogeneous hyperbolic systems (for example, shallow water equations (one layer) [47], Ripa system [48], Euler equations with gravitation [49], the system of Magnetohydrodynamics (MHD) equations with gravitational source term [53]). In this work, we adapted the UCS scheme for hyperbolic balance laws with geometrical source terms to the two-layer model. The motivation behind using the two-layer model lies in its ability to capture more complex and realistic behaviors in certain scenarios compared to the traditional one-layer shallow water model. For instance, in situations where there are density differences between the two layers or when there are interactions between two distinct fluids, the two-layer approach becomes necessary to model such phenomena accurately. The obtained numerical results are in general satisfactory whenever the bottom topography function is either a constant function or a smooth one. However, for non-smooth waterbed functions, the UCS scheme fails to properly capture steady-state solutions, and oscillations and other numerical storms appear in the early computational steps. To ensure the validity of numerical solutions and maintain important steady-state constraints, special well-balanced discretizations need to be implemented. The SGM method, initially developed for data reconstruction within a Godunov-type method for solving shallow water equations with source terms, was successfully adapted to suit the two-layer model. Specifically, we discretized the lower layer water height based on the discretizations of the interface level function and the bottom topography function. The resulting UCS-SGM scheme represents an extension of the method previously presented for shallow water equations (one-layer flow) to address the complexities of two-layer rigid-lid flows while ensuring well-balanced characteristics.

The paper is structured in the following way: the reformulation of the two-layer shallow water equations with rigid-lid and variable topography is presented in Sect. 2; the UCS-SGM scheme is constructed in Sect. 3 and later implemented and validated in Sect. 4. Concluding remarks are discussed in Sect. 5.

2 Shallow Water Model

In this paper, we investigate the numerical study of the propagation of shallow water internal waves bounded between a flat rigid-lid and a variable topography. The internal waves propagate between two layers of immiscible superposed fluids with different densities. Reasonable oceanographic assumptions are considered on the nature of the fluids. In fact, both fluids are considered to be homogeneous, incompressible, and inviscid. In this section, starting from the free surface two-layer shallow water equations, we formally recover the latter equations in the rigid-lid configuration.

By depth integrating the Euler equations and under the hydrostatic assumption, one gets the one-dimensional two-layer shallow water equations with a free surface:

$$\begin{cases} \partial_t h_1 + \partial_x \left(h_1 u_1\right) = 0, \\ \partial_t (h_1 u_1) + \partial_x \left(h_1 u_1^2 + \frac{1}{2} g h_1^2\right) = -g h_1 \partial_x h_2 - g h_1 \partial_x b, \\ \partial_t h_2 + \partial_x \left(h_2 u_2\right) = 0, \\ \partial_t (h_2 u_2) + \partial_x \left(h_2 u_2^2 + \frac{1}{2} g h_2^2\right) = -\gamma g h_2 \partial_x h_1 - g h_2 \partial_x b, \end{cases} \tag{1}$$

where $h_1, h_2 > 0$ denotes, respectively, the heights of the upper and lower layers, thus $h_1(t,x) = d_1 + \eta(t,x) - \zeta(t,x)$ and $h_2(t,x) = d_2 + \zeta(t,x) - b(x)$ with $\eta(t,x)$, $\zeta(t,x)$ and $b(x)$ are functions that represent, respectively, the surface, interface,

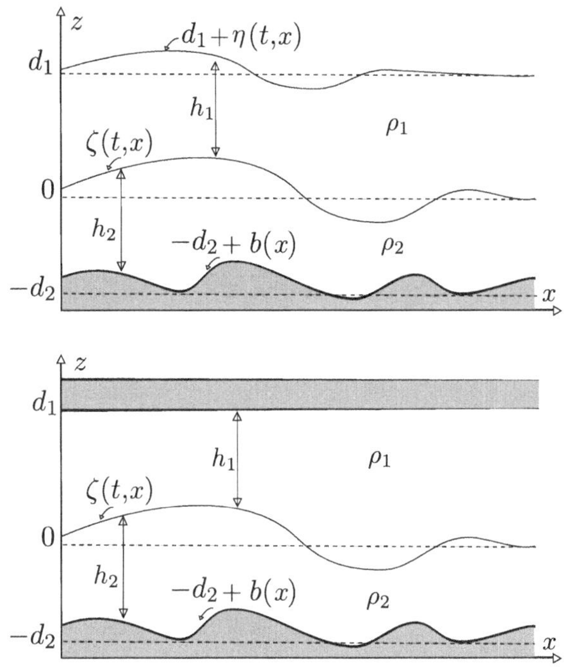

Fig. 1 Free surface configuration (top), rigid-lid configuration (bottom)

and bottom elevation from their rest position (see Fig. 1). The upper and lower fluids horizontal velocities are, respectively, denoted by u_1 and u_2. The density contrast between the two fluids is assumed to be small and thus we define

$$\gamma = \frac{\rho_1}{\rho_2} \in (0, 1)$$

as the density ratio between the two fluids. The acceleration of gravity is denoted by g. System (1) was originally described in [45] in the flat topography case and for a detailed derivation in the variable topography case we refer to [11] among others.

Several difficulties arise in the numerical computation of system (1). First, due to the hydrostatic pressure integration within the layers, the system contains non-conservative product terms $h_1\partial_x h_2$ and $h_2\partial_x h_1$ coupling the two-layer momentum equations. The coupling and the corresponding 4×4 flux matrix do not provide explicit eigenvalues [1, 32]. Actually, approximate eigenvalue expressions predict a loss of hyperbolicity whenever $u_1 - u_2$ is large [12]. Additionally, the discretization of the coupling terms is an extra challenge in the development of valid and robust numerical methods. Furthermore, the non-conservative source terms accounting for the variable topography make the construction of a well-balanced numerical scheme absolutely necessary in order to preserve the steady-state solutions [11, 13].

At this point, in order to simplify the setting, we assume that the interface is delimited from above by a flat rigid lid, thus one can neglect the displacement of the surface. This assumption is commonly used in the literature and reduces significantly the complexity in the model. Setting $\eta(t, x) = 0$, the height of the upper layer becomes $h_1 = d_1 - \zeta(t, x)$ and thus $\partial_t h_1 = -\partial_t h_2 = -\partial_t \zeta$. At this stage, a new variable is defined called the "shear mean velocity" variable v coupling the upper and lower layer horizontal velocities u_1 and u_2, see [22, 23] for more details:

$$v \equiv u_2 - \gamma u_1. \tag{2}$$

It is clear now that the first and third equations of (1) are equivalent, and lead to the same equation:

$$\partial_t \zeta + \partial_x \Big(\frac{h_1 h_2}{h_1 + \gamma h_2} v \Big) = 0.$$

A straightforward computation consisting of multiplying the second equation of (1) by γ and dividing it by $h_1 > 0$ with the use of the identity $\partial_t h_1 = -\partial_x (h_1 u_1)$ (coming from the mass equation on the upper layer) yields

$$\gamma \partial_t u_1 + \gamma u_1 \partial_x u_1 + \gamma g \partial_x (h_1 + h_2 + b) = 0. \tag{3}$$

Dividing the fourth equation of (1) by $h_2 > 0$ and using the identity $\partial_t (h_2) = -\partial_x (h_2 u_2)$ (coming from the mass equation on the lower layer) yields

$$\partial_t u_2 + u_2 \partial_x u_2 + g \partial_x (h_2 + \gamma h_1 + b) = 0. \tag{4}$$

Now, subtracting (3) from (4) and using the identity $h_2(t, x) + b(x) = d_2 + \zeta(t, x)$ yields

$$\partial_t v + g(1 - \gamma) \partial_x \zeta + \frac{1}{2} \partial_x \Big(\frac{h_1^2 - \gamma h_2^2}{(h_1 + \gamma h_2)^2} v^2 \Big) = 0.$$

Consequently, the shallow water model is written as a system of two conservative equations coupling two variables ζ and v:

$$\begin{cases} \partial_t \zeta + \partial_x \big(a(h_1, h_2) v \big) \; = \; 0, \\ \partial_t v \; + \; g(1 - \gamma) \partial_x \zeta \; + \; \partial_x \Big(c(h_1, h_2) v^2 \Big) \; = 0, \end{cases} \tag{5}$$

where we denote $a(h_1, h_2) = \dfrac{h_1 h_2}{h_1 + \gamma h_2}$, $c(h_1, h_2) = \dfrac{h_1^2 - \gamma h_2^2}{2(h_1 + \gamma h_2)^2}$, where $h_1 = d_1 - \zeta$ and $h_2 = d_2 + \zeta - b$ are, respectively, the heights of the top and low fluids.

The two-layer SWE with a rigid lid in a variable topography configuration (5) has a conservative structure without source terms which make them suitable for the implementation of a numerical scheme preserving the steady-state solutions without resorting to the well-balancing technique. Moreover, the eigenvalue expressions can be explicitly obtained. However, the SWE formulation (5) coupling the two conservative variables (ζ, v) is not suitable to the application of the surface gradient method (SGM) in the data reconstruction step. In fact, data reconstruction method based on the conservative variable ζ fails to reproduce accurate values because the interface gradient does not reproduce the real variation. To this end, we will reformulate in the next section the SWE system into an equivalent system of three evolution equations on h_1, h_2, and v. This reformulation is well adapted to the application of the SGM since the effects of topography slope are included and thus the steady-state solutions are preserved at the discrete level. Another possibility is to solve the SWE system without computing both heights of the top and low fluids, but only compute one of them and force the rigid-lid hypothesis. In fact, once we have one of the fluid heights, the other is uniquely determined. In this paper, we will show that even when computing both heights the SGM preserves the expected rigid-lid hypothesis.

Remark 1 The SWE system (5) is demonstrated to be well-posed in [29] in the case of a flat bottom in dimensions $d = 1$ and $d = 2$. We believe a similar result could be obtained in the variable-bottom configuration following the same techniques (energy-estimate methods for quasi-linear hyperbolic systems). Many authors have also derived and proved the system to be consistent, one can see [3, 7, 15, 19, 23].

2.1 *Reformulation of the Shallow Water Equations*

In the present section, the SWE system (5) is reformulated into a new system of three evolution equations on h_1, h_2, and v.

One just has to remark for the first equation that $\partial_t \zeta = -\partial_t h_1 = \partial_t h_2$, thus

$$\partial_t h_1 - \partial_x(a(h_1, h_2)v) = 0 \text{ and } \partial_t h_2 + \partial_x(a(h_1, h_2)v) = 0.$$

For the second equation, one just has to remark that $\partial_x \zeta = -\partial_x(h_1)$, we thus get

$$\partial_t v - g(1-\gamma)\partial_x(h_1) + \partial_x\Big(c(h_1, h_2)v^2\Big) = 0.$$

The system (5) can therefore be written in (h_1, h_2, v) variables as

$$\partial_t \mathbf{u} + \partial_x(f(\mathbf{u})) = 0, \tag{6}$$

with the following conservative variable and flux:

$$\mathbf{u} = \begin{pmatrix} h_1 \\ h_2 \\ v \end{pmatrix}, \quad f(\mathbf{u}) = \begin{pmatrix} -a(h_1, h_2)v \\ a(h_1, h_2)v \\ c(h_1, h_2)v^2 - g(1-\gamma)(h_1) \end{pmatrix}. \tag{7}$$

In the rest of the paper, we will assume that $d_1 = d_2 = 1$. A simple computation shows that the homogeneous system (6) is strictly hyperbolic provided that:

$$\begin{cases} \inf_{x\in\mathbb{R}} h_1 > 0, \\ \inf_{x\in\mathbb{R}} h_2 > 0, \\ \inf_{x\in\mathbb{R}} \left[g(1-\gamma) - \frac{\gamma(h_1+h_2)^2}{(h_1+\gamma h_2)^3} v^2 \right] > 0. \end{cases} \tag{8}$$

Two conditions amount to saying that both fluid layers remain of positive depth. Such a condition has to be imposed in the classical Saint-Venant (shallow water) system for surface waves. As a matter of fact, these two conditions simply consist in assuming that the deformation of the interface is not too large. The third condition has no counterpart for the Saint-Venant system. It can be viewed as a "trace" of the Kelvin–Helmholtz instabilities inherent to the full system. This condition consists of imposing a smallness assumption on v.

The Jacobian matrix of f has three eigenvalues:

$$\begin{aligned} \lambda_1 &= 0, \\ \lambda_2 &= l(\mathbf{u}) + \sqrt{a(h_1, h_2) m(\mathbf{u})}, \\ \lambda_3 &= l(\mathbf{u}) - \sqrt{a(h_1, h_2) m(\mathbf{u})}, \end{aligned} \tag{9}$$

where $l(\mathbf{u}) = \frac{h_1^2 - \gamma h_2^2}{(h_1+\gamma h_2)^2} v$ and $m(\mathbf{u}) = g(1-\gamma) - \frac{\gamma(h_1+h_2)^2}{(h_1+\gamma h_2)^3} v^2$.

Evidently, depending on the values of the density ratio γ, the eigenvalues (9) may become complex. In this case, the system is not hyperbolic and yields to the so-called Kelvin–Helmholtz instability at the interface separating the two layers. For the numerical simulations, we have in mind we consider two values of density ratios, $\gamma = 0.98$ and $\gamma = 0.7$. Nevertheless, it is worth mentioning at this stage that our numerical scheme is able to handle the more difficult tests associated to an aspect ratio closer to 1.

2.2 Steady-State Solutions

The steady-state solutions of shallow water models have attracted a lot of interest in the literature. Many efforts have been made to construct efficient numerical schemes that are able to preserve the steady states or at least a class of steady states.

Such schemes are commonly called well balanced. In fact, shallow water models with free surface contain a source term accounting for the variable topography. The well-balanced property is achieved with the help of a special discretization of this geometric source term, thus balancing the other terms in the flux [11, 13]. Here, we confine our work to the rigid-lid configuration where the two-layer shallow water model is a system of conservation law (i.e., no source terms) thus no well balancing is needed. The two-layer shallow water system with rigid lid (6)–(7) features steady-state solutions governed by the following system:

$$\partial_x(a(h_1, h_2)v) = 0,$$

$$\partial_x\Big(c(h_1, h_2)v^2 + g(1-\gamma)(1-h_1)\Big) = 0.$$

One can easily deduce that the steady states are characterized by

$$a(h_1, h_2)v = \text{constant},$$

$$c(h_1, h_2)v^2 + g(1-\gamma)(1-h_1) = \text{constant}.$$

It is very difficult to restore all steady states, in particular, the ones with non-vanishing velocity. In this work, we construct a numerical scheme that is able to preserve the lake-at-rest steady state defined by vanishing the velocity. In fact, most real-life problems reside in a flow regime slightly deviating from these steady states [4, 8]. The homogeneous rigid-lid system (6)–(7) features steady-state solutions that correspond to a lake-at-rest-type problem given by

$$\begin{cases} h_1 = K_1 = \text{constant}, \\ h_2 + b = K_2 = \text{constant}, \\ v = 0. \end{cases} \tag{10}$$

The two-layer shallow water system with rigid-lid formulation (6)–(7) is well adapted to the application of the SGM within the proposed finite volume UCS, hence preserving steady-state requirements at the discrete level.

3 Central Finite Volume Schemes for the Two-Layer Rigid-Lid Model

This section is about constructing a new finite volume scheme for the two-layer rigid-lid system (6)–(7) in one dimension. The proposed scheme extends the previously developed unstaggered central scheme (UCS) for hyperbolic systems of conservation laws to the case of rigid-lid systems; furthermore, and in order to fulfill the lake-at-rest steady-state requirement, we apply the surface gradient method and discretize the heavier fluid height in terms of its fluid level function. The resulting solution is shown to satisfy the lake-at-rest steady-state constraint in addition to the rigid-lid property up to machine accuracy.

3.1 One-Dimensional Unstaggered Central Scheme for the Homogeneous Rigid-Lid System

We consider the rigid-lid two-layer system (6) written in its conservative form in one spatial dimension

$$\begin{cases} \dfrac{\partial \mathbf{u}}{\partial t} + \dfrac{\partial f(\mathbf{u})}{\partial x} = 0, \\ u(x,0) = u_0(x). \end{cases} \tag{11}$$

Here $t > 0$ denotes the time variable and $x \in \Omega \subset \mathbb{R}$ denotes the spatial variable. $\mathbf{u}(x, t)$ denotes the unknown vector solution and $f(\mathbf{u})$ the flux defined in (7). We start by partitioning the computational domain Ω with the aid of Cartesian control cells C_i defined to be the subintervals $x_{i-1/2} \leq x \leq x_{i+1/2}$ centered at the nodes x_i with width $\Delta x = x_{i+1/2} - x_{i-1/2}$. We assume with no loss of generality that the numerical solution of the initial value problem (11) was already generated at time $t = t^n$ on the cells C_i, and we want to compute the numerical solution at time $t^{n+1} = t^n + \Delta t$. In the context of unstaggered central schemes, we generate the solution $\mathbf{u}(x, t)$ of system (11) using the piecewise linear interpolants $\mathcal{L}_i(x, t^n)$ defined on the cells C_i as follows:

$$\mathcal{L}_i(x, t^n) = \mathbf{u}_i^n + (x - x_i)(\mathbf{u}')_i^n \ \ \forall x \in C_i, \tag{12}$$

where $(\mathbf{u}')_i^n$ approximates $\partial \mathbf{u}(x_i, t^n)/\partial x$ to first order of accuracy. $\mathcal{L}_i(x, t^n)$ is used to approximate $\mathbf{u}(x, t)$ on the cell C_i, and only the cell-centered value is stored and is defined as

$$\mathbf{u}_i^n = \mathcal{L}_i(x_i, t^n) = \frac{1}{\Delta x} \int_{C_i} \mathcal{L}_i(x, t^n)\, dx. \tag{13}$$

The unstaggered central scheme [46] that we are discussing in this section develops gradually the numerical solution on a sole grid, but uses a second layer of cells $D_{i+1/2}$ defined by $x_i \leq x \leq x_{i+1}$ in order to bypass the resolution of the Riemann problems occurring at the cell interfaces. More precisely, we will generate first the numerical solution to system (11) at time t^{n+1} at the center of the cells $D_{i+1/2}$; this will produce the estimates $\mathbf{u}^{n+1}_{i+1/2} \approx \mathbf{u}(x_{i+1/2}, t^{n+1})$. A piecewise linear reconstruction of the cell-centered solution values is used to back-project $\mathbf{u}^{n+1}_{i+1/2}$ on the cells C_i and generate the solution $\mathbf{u}^{n+1}_i$ on the cells C_i. In summary, the unstaggered central scheme (UCS) is described as follows: Starting with the cell-centered numerical solution $\mathbf{u}^n_i$, we construct the piecewise linear interpolants $\mathcal{L}_i(x, t^n)$ and then we integrate the PDE in (11) on the rectangle $[x_i, x_{i+1}] \times [t^n, t^{n+1}]$, applying Green's theorem, we obtain

$$\mathbf{u}^{n+1}_{i+\frac{1}{2}} = \mathbf{u}^{n}_{i+\frac{1}{2}} - \frac{1}{\Delta x}\int_{t^n}^{t^{n+1}} \left[f\left(\mathbf{u}(x_{i+1}, t)\right) - f\left(\mathbf{u}(x_i, t)\right)\right] dt. \tag{14}$$

Here $\mathbf{u}^n_{i+1/2}$ is the interpolated numerical solution at the center of the intervals $[x_i, x_{i+1}]$; Taylor series are used to reconstruct the interpolants as follows:

$$\mathbf{u}^{n}_{i+\frac{1}{2}} = \frac{1}{2}\left(\mathbf{u}^n_i + \mathbf{u}^n_{i+1}\right) + \frac{\Delta x}{8}\left((\mathbf{u}')^n_i - (\mathbf{u}')^n_{i+1}\right). \tag{15}$$

Here $(\mathbf{u}')^n_i$ stands as a limited numerical partial derivative approximating $\frac{\partial}{\partial x}\mathbf{u}(x_i, t^n)$ and is obtained using the MC-θ limiter [50].

The flux integrals in Equation (14) are estimated (accurate to second order) with the help of the midpoint quadrature formula, we obtain

$$\mathbf{u}^{n+1}_{i+\frac{1}{2}} = \mathbf{u}^n_{i+1/2} - \frac{\Delta t}{\Delta x}\left[f(\mathbf{u}^{n+\frac{1}{2}}_{i+1}) - f(\mathbf{u}^{n+\frac{1}{2}}_{i})\right]. \tag{16}$$

Taylor series expansions in time will yield the expected solution values at time $t^{n+1/2}$ as follows. The required anticipated values at the transitional fractional time steps are reached with the aid of a first order in time in addition to

$$\mathbf{u}^{n+\frac{1}{2}}_i = \mathbf{u}^n_i - \frac{\Delta t}{2}(f')^n_i. \tag{17}$$

Here the conservation law in (11) was used to substitute the time derivative by the spatial derivative of the flux. Similar to Eq. (15), the term $(f')^n_i$ denotes the spatial numerical derivative of the flux function evaluated at $\mathbf{u}^n_i$; these numerical derivatives of the flux function can be obtained either by directly limiting the flux values, or else by applying the chain rule; the latter case requires a matrix multiplication of the Jacobian matrix $\partial f/\partial \mathbf{u}$ by $(\mathbf{u}')^n_i$ previously obtained and used in Eq. (15). To retrieve the numerical solution $\mathbf{u}^{n+1}_i$ at the next time-step t^{n+1} on C_i we apply a projection procedure of $\mathbf{u}^{n+1}_{i+\frac{1}{2}}$ back on the control cells C_i; this is based on Taylor series expansions in space; the resulting back-projection step takes its final form as

$$\mathbf{u}^{n+1}_i = \frac{1}{2}\left(\mathbf{u}^{n+1}_{i-\frac{1}{2}} + \mathbf{u}^{n+1}_{i+\frac{1}{2}}\right) + \frac{\Delta x}{8}\left((\mathbf{u}')^{n+1}_{i-\frac{1}{2}} - (\mathbf{u}')^{n+1}_{i+\frac{1}{2}}\right). \tag{18}$$

A limiting procedure is again applied in Eq. (18) to generate the spacial numerical derivatives of the solution obtained in the staggered dual cells.

3.2 *Surface Gradient Method for the Homogeneous Rigid-Lid System*

When the proposed numerical scheme is applied to rigid-lid problems on flat bottom topographies, the obtained numerical solution remains admissible; however, as we will see in Sect. 5, as soon as the waterbed/topography function becomes variable or non-smooth, nonphysical oscillations arise and numerical storms are observed in a motionless lake. This is similar to the case of a single-layer SWE systems, where some oscillations and other nonphysical waves were generated because of the ill-balanced discretization of the partial derivatives occurring in the source term as well as in the flux term of the balance-law system describing the shallow water equations.

In its current form, the constructed central finite volume method cannot handle properly steady-state solutions of the form (10) whenever applied to the rigid-lid two-layer flow systems and the equation $\mathbf{u}_i^{n+1} = \mathbf{u}_i^n$ doesn't hold, even if the initial conditions correspond to a steady-state problem. An additional treatment is required; mainly the forward projection step (15) at time t^n should be processed according to the surface gradient method (SGM), with a similar process for the backward projection step (18). In fact only the height function of the denser fluid should be projected (forth and back) according to the SGM approach. In this work, we will extend the methods discussed in [48, 49, 52] to the case of the rigid-lid, two-layer flow problems. The main idea consists of estimating $\partial h_2(x,t)/\partial x$ with the aid of the function $H_2(x,t) = h_2(x,t) + b(x)$; here the function $H_2(x,t)$ represents the denser fluid level function.

As in [18, 47, 52], we start by assuming that the waterbed function is pointwise defined at the cell endpoints $x_{i+1/2}$, i.e., we assume that the point values $b_{i+\frac{1}{2}}$ are given at the nodes $x_{i+\frac{1}{2}}$, and then we set the bottom elevation function values $b_i = \frac{1}{2}\left(b_{i+\frac{1}{2}} + b_{i-\frac{1}{2}}\right)$ at center of the cells C_i. The linear interpolants $b_i(x)$ estimating the waterbed function are then reconstructed on the control cells C_i, and we obtain

$$b_i(x) = b_i + \frac{b_{i+\frac{1}{2}} - b_{i-\frac{1}{2}}}{\Delta x}(x - x_i), \forall x \in C_i,$$

and the equations $b_i(x_i) = b(x_i) = b_i$ hold for all i. On the other hand, we apply spacial linear reconstruction techniques of the denser fluid height function $h_2(x,t)$ on the control cells; these reconstructions will be performed indirectly in terms of the linear reconstructions of the level functions $H_2(x,t) = h_2(x,t) + b(x)$.

The linearization of $H_2(x,t)$ on the cells C_i is performed with the aid of a limiting step of the numerical derivatives of denser fluid cell-centered elevation function values $(H_2)_i^n = (h_2)_i^n + b_i$. As for the linearization of $h_2(x,t)$, it is obtained using the numerical derivatives $(h_2')_i^n$; they are obtained as follows:

$$(h_2')_i^n = (H_2')_i^n - \frac{1}{\Delta x}(b_{i+1/2} - b_{i-1/2}). \tag{19}$$

The numerical derivatives obtained in Eq. (19) are used in the forward projection step (15) of the h_2−component in $\mathbf{u}_i^n$. Next, for the back-projection step of $\mathbf{u}_{i+1/2}^{n+1}$ on the control cells C_i, we apply once more the SGM approach. We set the denser fluid level function at the center of the staggered nodes to be $(\tilde{H}_2)_{i+1/2}^n$ with

$$(\tilde{H}_2)_{i+1/2}^{n+1} = (h_2)_{i+1/2}^{n+1} + \tilde{b}_{i+1/2}, \tag{20}$$

where $\tilde{b}_{i+1/2} = b_{i+1/2} - \frac{1}{2}\big(b_{i+1/2} - (b_i + b_{i+1})/2\big)$ is the corrected bottom elevation values. This correction is required, mainly because the bottom topography function $b_i(x)$ is linear on the original grid C_i only and not on the dual staggered cells.

Finally, we generate the numerical derivative $(h_2')_{i+1/2}^{n+1}$ using a limiting step of the numerical derivatives of the level function $(\tilde{H}_2)_{i+1/2}^{n+1}$:

$$(h_2')_{i+1/2}^{n+1} = (\tilde{H}_2{}')_{i+1/2}^{n+1} - \frac{1}{\Delta x}(b_{i+1} - b_i). \tag{21}$$

Equation (21) is only needed in the back-projection procedure (18) leading to the values of h_2 in $\mathbf{u}_i^{n+1}$. This brings us to the main theorem of this work.

Theorem 1 *If the numerical solution* $\mathbf{u}_i^{n+1}$ *of the rigid-lid two-fluid flow system* (6), (7) *is generated at time* $t = t^{n+1}$ *using the UCS method* (15)–(18), *and if the forth- and back-projections of the* h_2*-component in* $\mathbf{u}_i^n$ *and* $\mathbf{u}_{i+1/2}^{n+1}$ *are processed with the aid of Equations* (19) *and* (21), *then whenever* $\mathbf{u}_i^n$ *corresponds to a steady-state solution of the form* (10) *at time* t^n, *then* $\mathbf{u}_i^{n+1}$ *will remain as such at time* t^{n+1}.

Proof To establish the theorem we will show, component-wise, that $\mathbf{u}_i^{n+1} = \mathbf{u}_i^n$, for all i whenever $\mathbf{u}_i^n$ satisfies system (10) at the discrete level.

It is enough to prove that $(h_2)_i^{n+1} = (h_2)_i^n\ \forall i$ since only the h_2-component is treated according to the SGM. The h_1-component is updated using the original UCS scheme, see Remark 2.

When the forth- and back-projection steps of the h_2−component are processed following the SGM, we have

$$(h_2)_{i+1/2}^n = \frac{1}{2}\left((h_2)_i^n + (h_2)_{i+1}^n\right) + \frac{\Delta x}{8}\left[(h_2')_i^n - (h_2')_{i+1}^n\right]. \tag{22}$$

But since the denser fluid height function $(h_2)_i^n$ is linearized in terms of its elevation function $(H_2)_i^n = (h_2)_i^n + b_i$, we obtain

$$(h_2')_i^n = (H_2')_i^n - \frac{b_{i+1/2} - b_{i-1/2}}{\Delta x}.$$

Substituting the latter in (22), we obtain while recalling that $(H_2)_i^n = (h_2)_i^n + b_i = K_2, \forall i$:

$$(h_2)_{i+1/2}^n = \frac{1}{2}\left((h_2)_i^n + (h_2)_{i+1}^n\right) - \frac{1}{2}\left[b_{i+1/2} - \frac{b_i + b_{i+1}}{2}\right]. \quad (23)$$

In a similar way for the back-projection of the h_2-component, we apply the SGM, we obtain, knowing that $(\tilde{H}_2)_{i+1/2}^{n+1/2} = (h_2)_{i+1/2}^{n+1/2} + \tilde{b}_{i+1/2}$ is constant for all i,

$$(h_2)_i^{n+1} = \frac{1}{2}\left((h_2)_{i-1/2}^{n+1} + (h_2)_{i+1/2}^{n+1}\right) - \frac{1}{8}\left[-(b_i - b_{i-1}) + (b_{i+1} - b_i)\right] \quad (24)$$

Substituting Eq. (23) in (24) we obtain $(h_2)_i^{n+1} = (h_2)_i^n, \forall i$. □

Remark 2 In this work only the h_2−component of the numerical solution is treated according to the SGM. The h_1−component is updated using the original UCS scheme without any particular treatment; this doesn't interfere with the lake-at-rest constraint, and in the case of steady-state solutions, the requirement $(h_1)_i^{n+1} = (h_1)_i^n, \forall i$ is immediately satisfied as soon as $h_2 + b = K_2$ constraint is fulfilled at the discrete level. The main reason behind this is the fact that the numerical base scheme (UCS) is capable of updating a constant (or a zero) solution exactly.

Remark 3 The developed well-balanced finite volume scheme follows the same stability requirement (CFL condition) as the original staggered Lax–Friedrichs method and also the Nessyahu and Tadmor finite volume method [42]; the numerical treatment of the source term along with the SGM treatment doesn't restrict the CFL condition. The numerical experiments we report in the next section are processed with a CFL number of 0.485. The time stepping Δt is generated at the beginning of each iteration with the aid of the eigenvalues $\{\lambda_i\}_{i=1}^3$ of the Jacobian matrix $\partial f/\partial \mathbf{u}$ using the equation $\max_{1\le i\le 3}(|\lambda_i|)\frac{\Delta t}{\Delta x} = 0.485$.

4 Numerical Experiments

In this section, we implement the derived finite volume methods and compute the numerical solution of some selected two-layer rigid-lid problems set on smooth or discontinuous variable topographies. We report the findings of the simulations, validate the order of convergence of the scheme, and compute the steady state whenever relevant. In all the numerical tests, we set the gravitational constant $g = 9.81\ m/s^2$.

Table 1 Steady-state problem on a continuous waterbed: L_1-error and table of convergence

n	ℓ_1-error: $h_1 + h_2 + b$	Order
32	9.212×10^{-10}	–
64	2.274×10^{-10}	2.018
128	5.294×10^{-11}	2.103

4.1 Test Case 1: Steady-State Problem on a Continuous Waterbed

Our first numerical test case features the problem of a lake at rest set on a smooth waterbed function. The bottom topography function is given by $b(x) = e^{-x^2}$ and the initial conditions are given by

$$(h_1(x, 0), h_2(x, 0), v(x, 0)) = (2 - h_2(x, 0) - b(x), 1.5 - b(x), 0).$$

Furthermore, we set $\gamma = 0.98$. The computational domain spans a length of 10 m, where $x \in \Omega = [-5, 5]$ and the numerical solution is generated at the final time $t_f = 8$ s using both UCS and UCS-SGM methods. The obtained results are reported in Figs. 2 and 3. Figure 2 shows the profile of the fluids' elevation functions in addition to the bottom topography at the initial time (top Figure) and at the final time (bottom Figure) obtained using the UCS scheme. The observed numerical results show that the UCS scheme fails to satisfy the lake-at-rest property and generates nonphysical numerical waves since the early computational steps. Next we solve this same problem using the UCS-SGM scheme and we report the findings in Fig. 3. The fluids' elevation functions presented in Fig. 3 (top) satisfy the lake-at-rest property at the discrete level and up to machine accuracy. The nonphysical waves observed in Fig. 2 (bottom) disappeared as shown in Fig. 3 (top) and the velocity $v(x, t)$ remains within a range of 10^{-17}. Next we calculate ℓ_1-error for $h_1 + h_2 + b$ for $n = 32, 64$, and 128 grid points and we compute the corresponding order of convergence. The obtained results reported in Table 1 endorse the second order of convergence of the UCS-SGM method whenever the solution is a smooth function. The ℓ_1-error for $h_1 + h_2 + b$ as a function of the number of grid points n is plotted in a log-log scale in Fig. 4. The convergence rate is indeed quadratic. Finally, we confirm the lake-at-rest/rigid-lid property of the numerical solution at the discrete level and calculate the ℓ_∞-errors for $h_1 + h_2 + b, h_2 + b$, and v on 200 grid points. The obtained results presented in Table 2 (first row) show that the computed solution fulfills the steady-state requirement at the discrete level and up to machine accuracy.

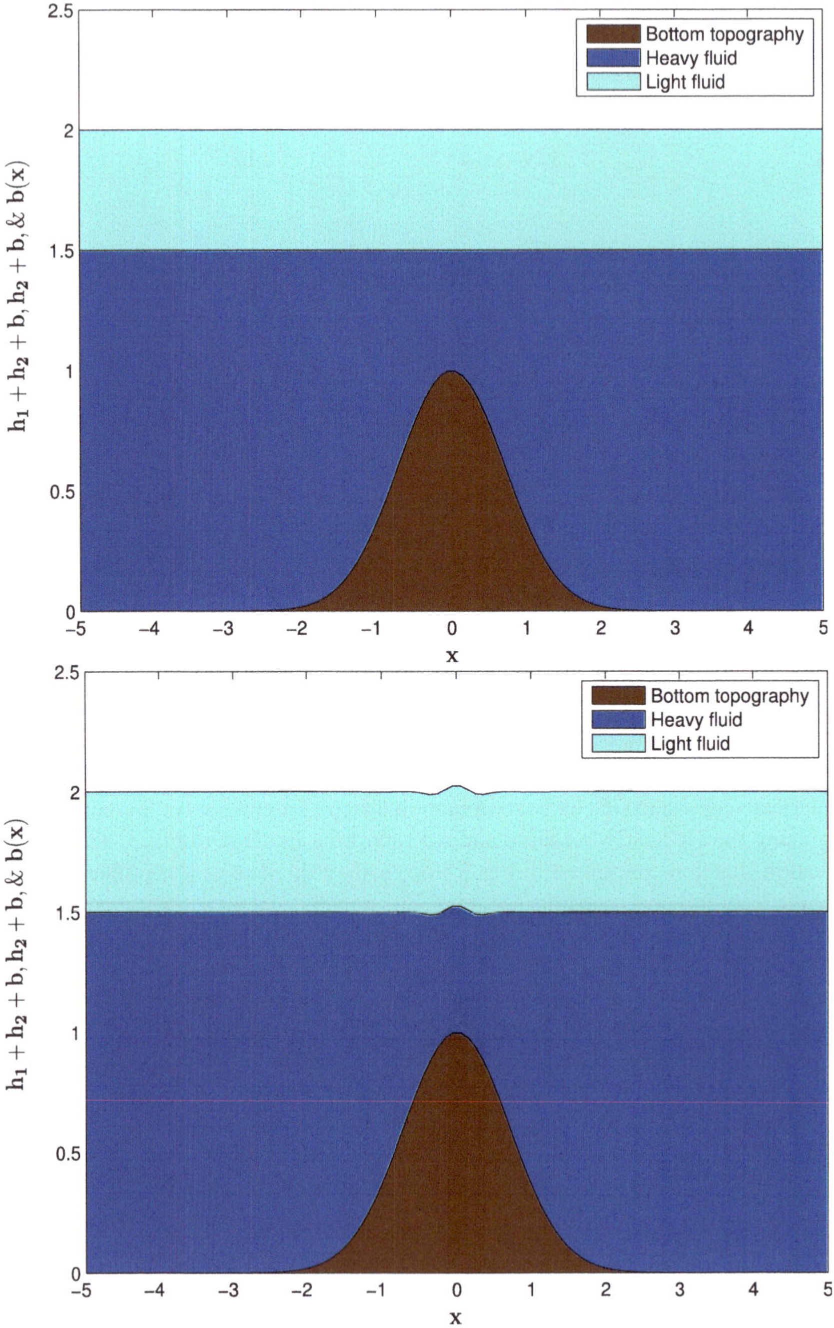

Fig. 2 Test case 1: Profile of the initial data (top) and the numerical solution obtained using the UCS scheme (bottom). The numerical base scheme fails to maintain the steady state at the discrete level and generates numerical storms since the early steps of computations

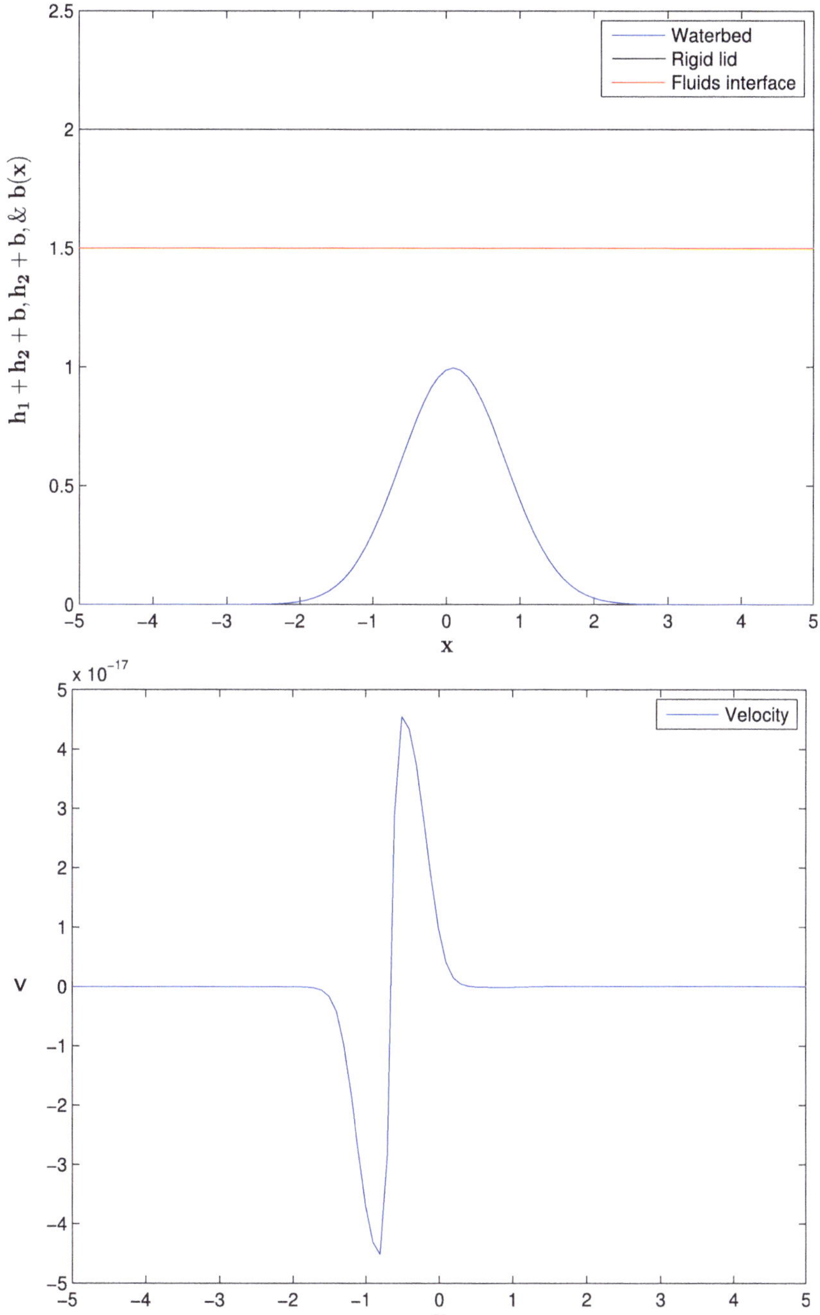

Fig. 3 Test case 1: Fluids' level functions (top) and velocity (bottom) obtained using the UCS-SGM scheme at the final time $t_f = 8$ s. The steady state remains fulfilled at the discrete level up to machine accuracy

Table 2 ℓ_∞-errors for $h_1 + h_2 + b$, $h_2 + b$, and v in the numerical solutions obtained using the UCS-SGM scheme for test cases 1 and 2

Test case	ℓ_∞-error: $h_2 + b$	ℓ_∞-error: $h_1 + h_2 + b$	ℓ_∞-error: v
1	4.8985×10^{-12}	4.8987×10^{-12}	2.4627×10^{-17}
2	0	0	0

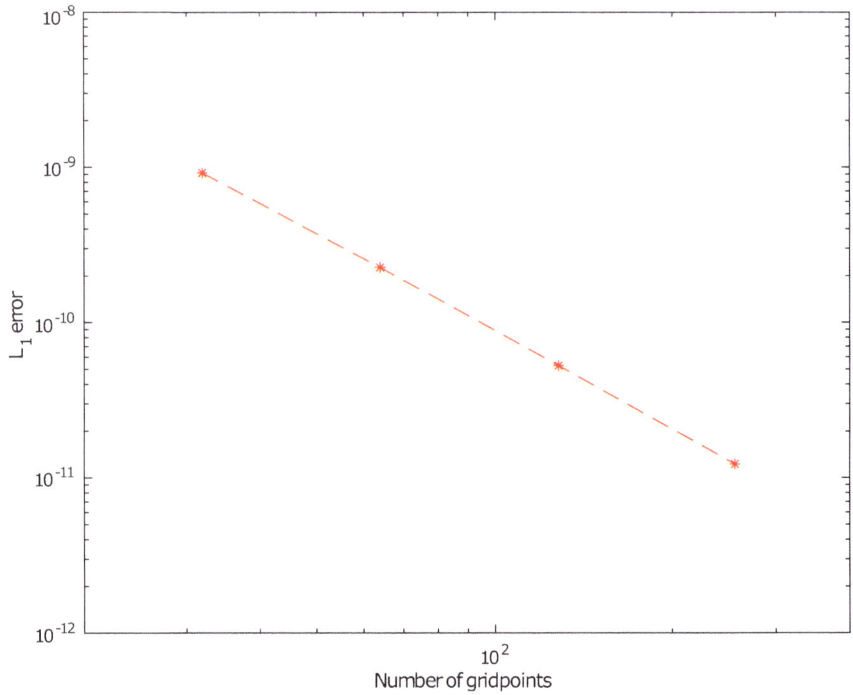

Fig. 4 ℓ_1 errors as a function of n (log-log plot)

4.2 *Test Case 2: Steady-State Problem on a Discontinuous Waterbed*

We consider for our second numerical test case a steady-state problem on a discontinuous waterbed. We set the interval $\Omega = [0, 10\text{ m}]$ to be the computational domain and we define the waterbed as

$$b(x) = \begin{cases} 0.5, & \text{if } |x-5| \leq 1 \\ 0, & \text{otherwise.} \end{cases}$$

This problem is a generalization of a shallow water equations problem, previously considered in [18, 47], to the case of two-layer rigid-lid systems. The initial conditions for this test case correspond to a steady-state problem verifying the conditions in (10). We set $\gamma = 0.98$ and we consider the initial conditions

$$\mathbf{u}_0(x) = \mathbf{u}(x, t = 0) = \begin{pmatrix} h_1(x, 0) \\ h_2(x, 0) \\ v(x, 0) \end{pmatrix} = \begin{pmatrix} 2 - h_2(x, 0) - b(x) \\ 1 - b(x) \\ 0 \end{pmatrix}. \quad (25)$$

Variants of this problem were considered in the setup of SWE systems in several papers such as in [2, 6, 10, 25, 27, 51, 52]. The analytic (exact) solution to this test case remains identical to the initial condition for any $t > 0$, i.e., $\mathbf{u}(x, t) = \mathbf{u}(x, t = 0) = \mathbf{u}_0(x), \forall t > 0, \forall x \in \Omega$. The numerical solution to this problem is computed using both the UCS and UCS-SGM schemes. The obtained results are reported in Figs. 5 and 6. Figure 5 (top) shows the initial fluids' elevation functions along with the discontinuous waterbed while Fig. 5 (bottom) shows the two fluids' level functions obtained at time $t_f = 0.4$ s on 200 grid points using the UCS scheme. The reported results show that the UCS scheme fails to maintain the steady-state solution around the jumps of the waterbed and generates nonphysical waves since the early steps of computations. These oscillations are felt by the lighter fluid and the rigid-lid constraint is breached by the numerical solution generated using the UCS scheme. The solution obtained using the UCS-SGM method is reported in Fig. 6; we can see clearly that the steady state remains satisfied throughout the computations and the numerical solution remains identical to the initial condition, thus endorsing the potential of the proposed UCS-SGM scheme to handle properly two-layer rigid-lid problems. As we did for test case 1, here also we confirm the lake-at-rest/rigid-lid property of the numerical solution at the discrete level and calculate the ℓ_∞-errors for $h_1 + h_2 + b$, $h_2 + b$, and v on 200 grid points. The obtained results presented in Table 2 (second row) show that the computed solution fulfills the steady-state requirement at the discrete level exactly.

4.3 Test Case 3: Dam Break Test Case on a Discontinuous Topography

We consider for our next numerical experiment a dam break problem set over a discontinuous waterbed topography. This test case is an extension of a similar shallow water equation problem, previously considered in [47], to the case of rigid-lid, two-layer flow system. Variants of this problem were considered in the setup of SWE systems in several papers such as [25, 27, 51, 52]. The computational domain spans a length of 2 m, where $x \in \Omega = [-1, 1]$ and the waterbed's topography features a discontinuous bump given by

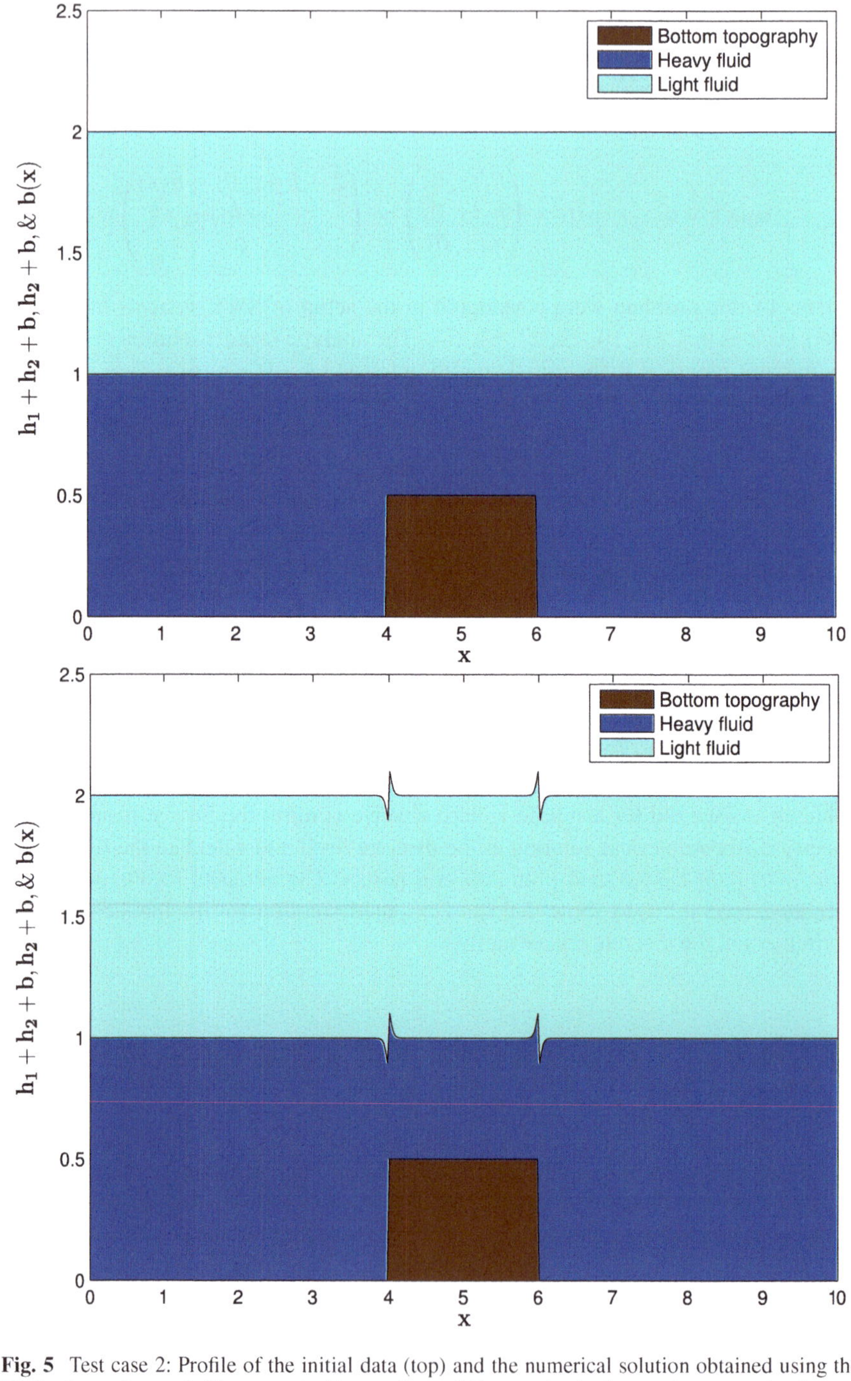

Fig. 5 Test case 2: Profile of the initial data (top) and the numerical solution obtained using the UCS scheme (bottom). The numerical base scheme fails to maintain the steady state at the discrete level and generates numerical storms since the early steps of computations

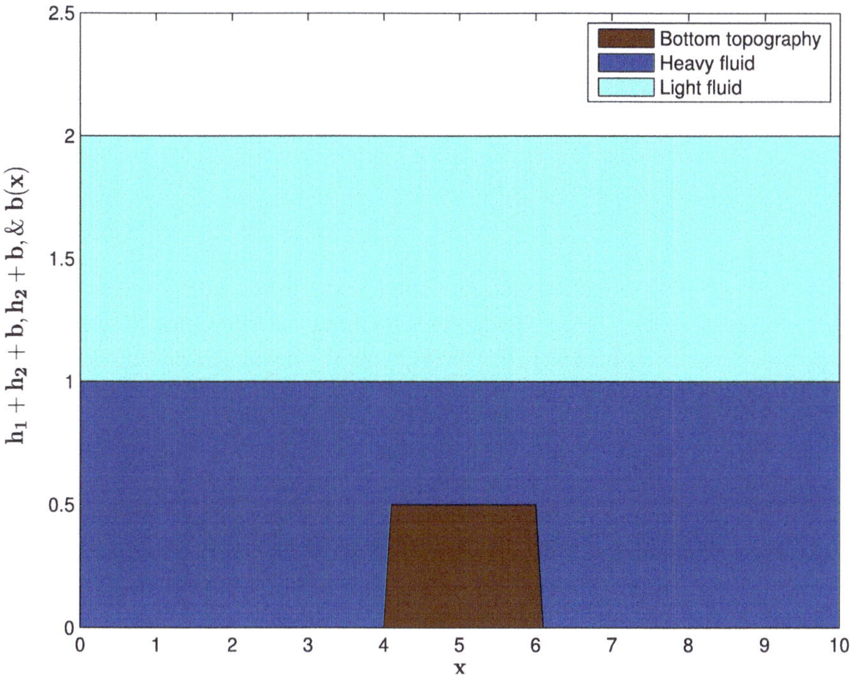

Fig. 6 Test case 2: Fluid's elevation functions obtained at time $t_f = 4$ s with the aid of the UCS-SGM method. The steady state is exactly maintained throughout the computations

$$b(x) = \begin{cases} 0.5, & \text{if } |x| < 0.6 \\ 0, & \text{otherwise.} \end{cases}$$

The initial data for $\mathbf{u}(x, 0) = (h_1, h_2, v)$ is as follows:

$$\mathbf{u}(x, 0) = \begin{cases} (2 - h_2(x, 0) - b(x), 1 - b(x), 0), & \text{if } x < 0 \\ (2 - h_2(x, 0) - b(x), 0.8 - b(x), 0), & \text{otherwise.} \end{cases}$$

In addition, we set $\gamma = 0.7$. The numerical solution of the two-layer rigid-lid system is generated at the final time $t_f = 0.4$ s with the aid of the UCS and the UCS-SGM methods on m grid points with $m \in \{100, 200, 400\}$. Figure 7 (top) shows the elevation functions of the two-layer system obtained at the final time with the aid of the UCS method. Oscillations in the fluid height functions h_1 and h_2 are observed at the boundaries of the rectangular bump and are due to the forward (15) and backward (18) projections in the original UCS scheme. These oscillations disappear completely when the surface gradient method is embedded within the UCS method as shown in Fig. 7 (bottom) where we show the fluids' level functions obtained with the aid of the UCS-SGM scheme.

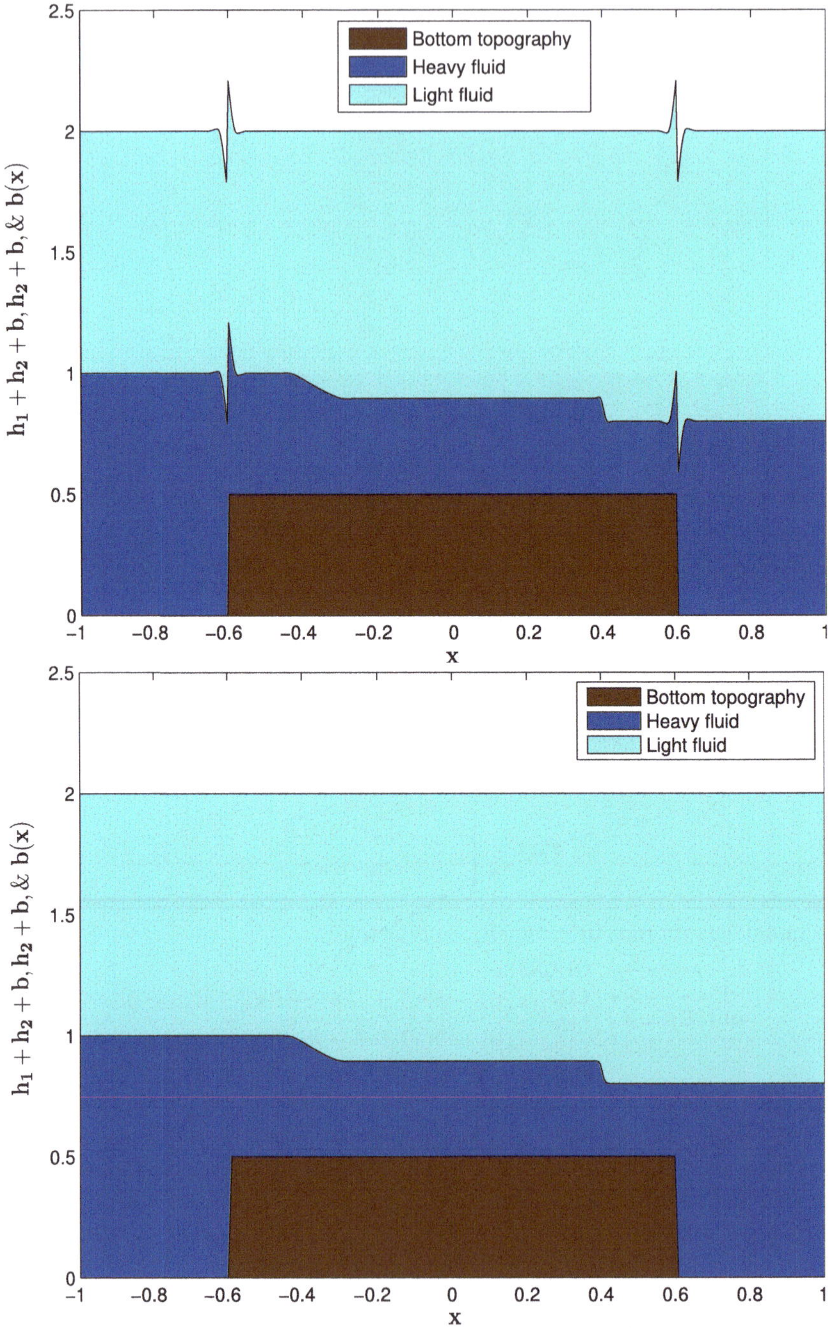

Fig. 7 Dam break on a discontinuous waterbed (rectangular bump): Two-layer fluid elevation functions obtained at time $t_f = 0.4$s using the UCS (top) and UCS-SGM (bottom) schemes

A grid convergence for both schemes (UCS and UCS-SGM) is performed and the numerical solution is generated at the final time on 10^2 (o-curve), 2×10^2 (x-curve), and 4×10^2 (solid line curve) grid points. The profile of the two-layer fluid elevation functions obtained using the UCS scheme and the UCS-SGM scheme is reported in Figs. 8 (top) and (bottom), respectively. Furthermore, the numerical solution obtained using the UCS-SGM satisfies the rigid-lid constraint $(h_1 + h_2 + b(x) = 2)$ at the discrete level up to machine accuracy.

4.4 Test Case 4: Dam Break Problem on a Smooth Waterbed Topography

Our next numerical experiment features a dam break problem set on a smooth waterbed. The waterbed's topography is modeled by the function $b(x) = e^{-x^2}$, and the computational domain spans a length of 6 m, where $x \in \Omega = [-3, 3]$. The initial data is summarized as follows:

$$\mathbf{u}(x, t = 0) = \begin{cases} (2 - h_2(x, 0) - b(x), 1.25 - b(x), 0), & \text{if } x \leq 0, \\ (2 - h_2(x, 0) - b(x), 1.75 - b(x), 0), & \text{otherwise.} \end{cases}$$

We set $\gamma = 0.98$. We implement the derived finite volume methods and we generate the numerical solution at time $t_f = 8$ s with the aid of the UCS and UCS-SGM schemes. The generated numerical solution is reported in Figs. 9, 10 and 11.

Figure 9 illustrates the profile of the elevation functions of the two-layer fluids at time $t_0 = 0$. Figure 10 shows the fluids' elevation functions (level functions) in the numerical solution generated at time $t_f = 8$ s, obtained using the UCS scheme on 200 grid points. The observed results show that the UCS scheme can still capture acceptable results in the case of unsteady flows on smooth variable-bottom topographies, but wrinkles in the computed rigid lid occur around $x = 0$ and propagate throughout the computational domain. When the UCS-SGM scheme is applied to solve the initial value problem the wrinkles in the computed rigid lid disappear as shown in Fig. 11 where we present the fluids' elevations at the final time obtained on 200 grid points. The ℓ_∞-norm of the error in the computed rigid-lid remains within 10^{-16} when the UCS-SGM is implemented. Finally, Fig. 12 shows a comparison between the solutions computed using the UCS-SGM scheme and the solutions obtained using a Riemann-based numerical method. A Harten, Lax, and van Leer (HLL) approximate Riemann solver is used, with a monotonic upstream scheme for conservation laws (MUSCL). This second-order discretization in space is associated to a second-order Runge–Kutta method (RK2) discretization in time. As a result, one gets our previously developed MUSCL-RK2 scheme, see [36] for more details. The computed fluids' elevation functions using the UCS-SGM scheme (red color) are compared to the results obtained using the MUSCL-RK2 scheme (green color) with 200 grid points; a rarefaction wave propagating to the left boundary and a shockwave

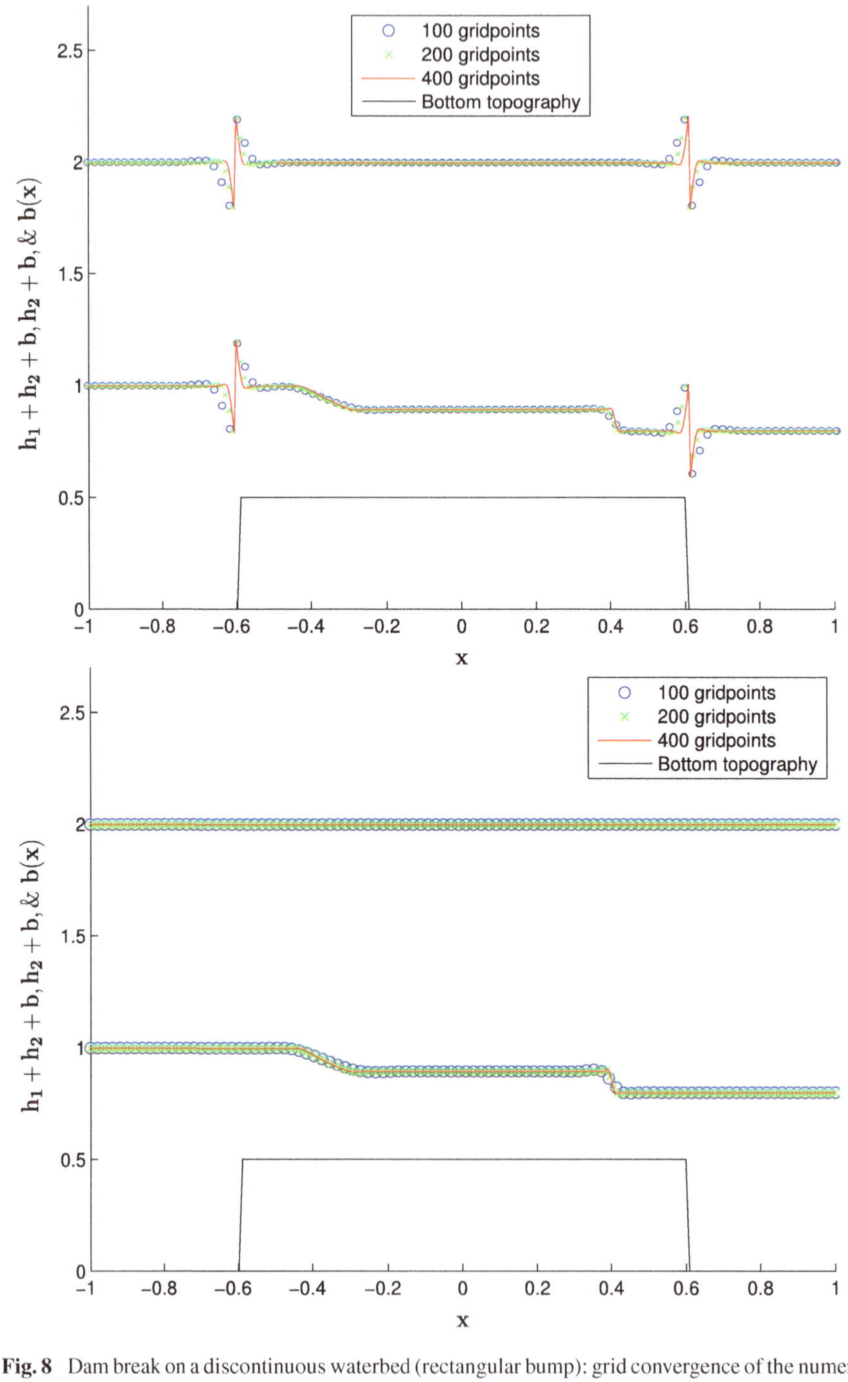

Fig. 8 Dam break on a discontinuous waterbed (rectangular bump): grid convergence of the numerical solutions obtained using the UCS (top) and the UCS-SGM schemes (bottom), on 100, 200, and 400 grid points

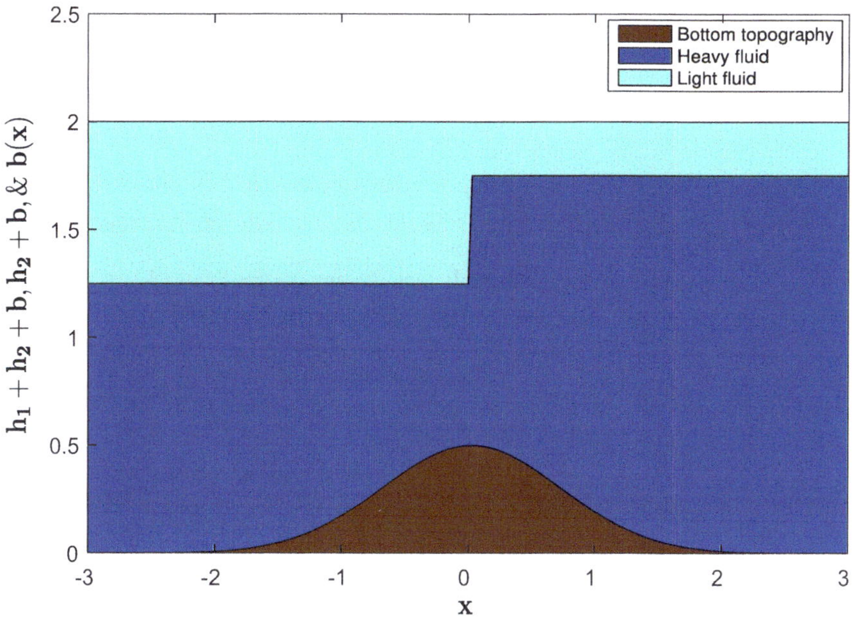

Fig. 9 Test case 4: Fluids' elevation functions at the initial time in addition to the waterbed topography

propagating to the right boundary are observed for $t \in \{2.5, 4, 6, 8\text{ s}\}$. A very good agreement is observed between the two numerical solutions at different times.

4.5 *Test Case 5: Dam Break Problem on a Discontinuous Waterbed*

Our last test case is about a dam break problem set on a discontinuous bottom topography (step function). The initial data features a discontinuity at the point $x_0 = 0$ of the computational domain $\Omega = [-3, 3]$ with a 6 m length. The left and right states across the discontinuity for $\mathbf{u} = (h_1, h_2, v)$ are $\mathbf{u}_l = (2 - h_2(x, 0) - b(x), 0.75 - b(x), 0)$ and $\mathbf{u}_r = (2 - h_2(x, 0) - b(x), 1.25 - b(x), 0)$. The bottom topography function is defined by

$$b(x) = \begin{cases} 0 & \text{if } x < 0, \\ 0.5 & \text{otherwise.} \end{cases}$$

We set $\gamma = 0.98$. The numerical solution is generated at time $t_f = 8$ s with the aid of both the UCS and the UCS-SGM schemes. The obtained numerical results are presented in Figs. 13 and 14.

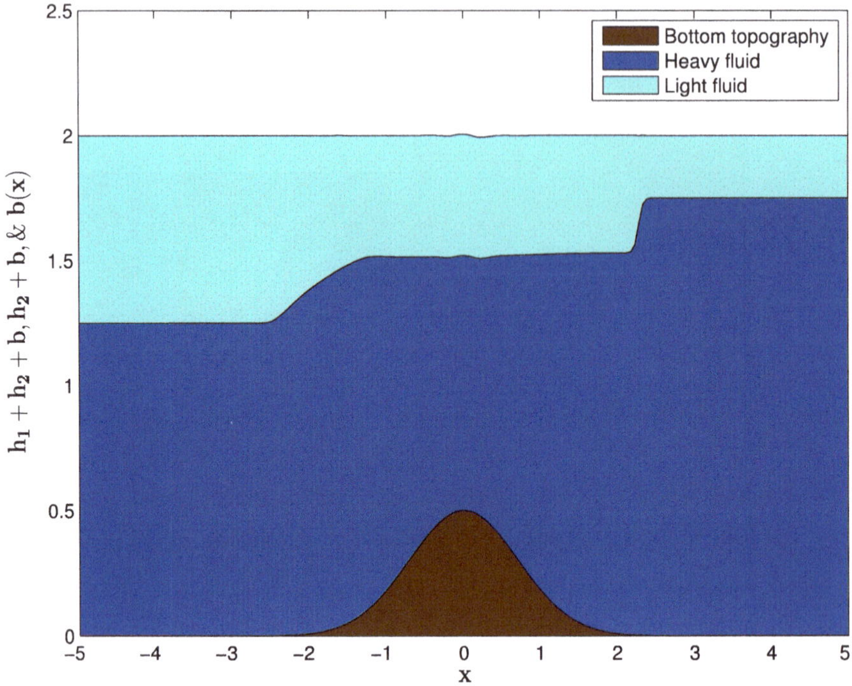

Fig. 10 Test case 4: Computed fluids' elevation functions using the UCS scheme. The computed rigid lid shows wrinkles around the point $x = 0$

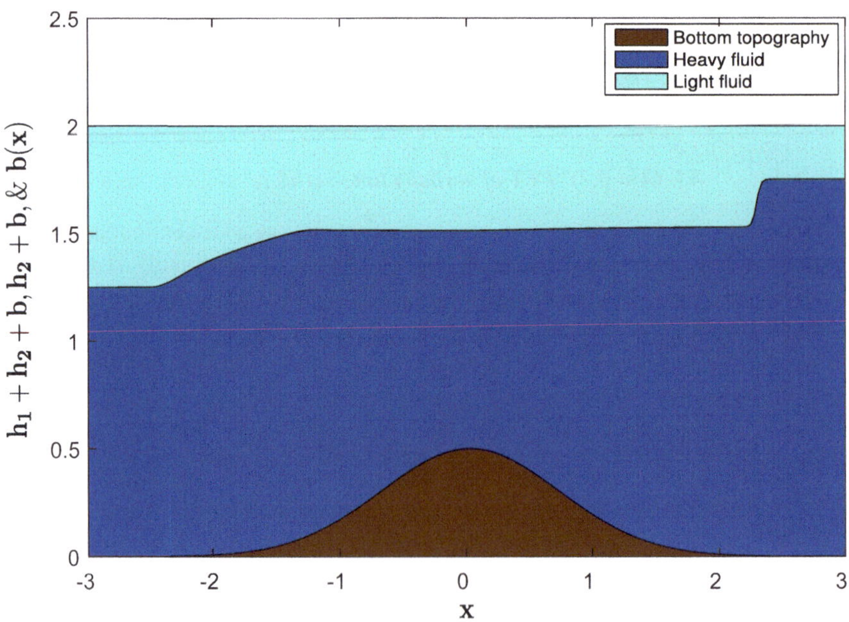

Fig. 11 Test case 4: Computed fluids' elevation functions using the UCS-SGM scheme

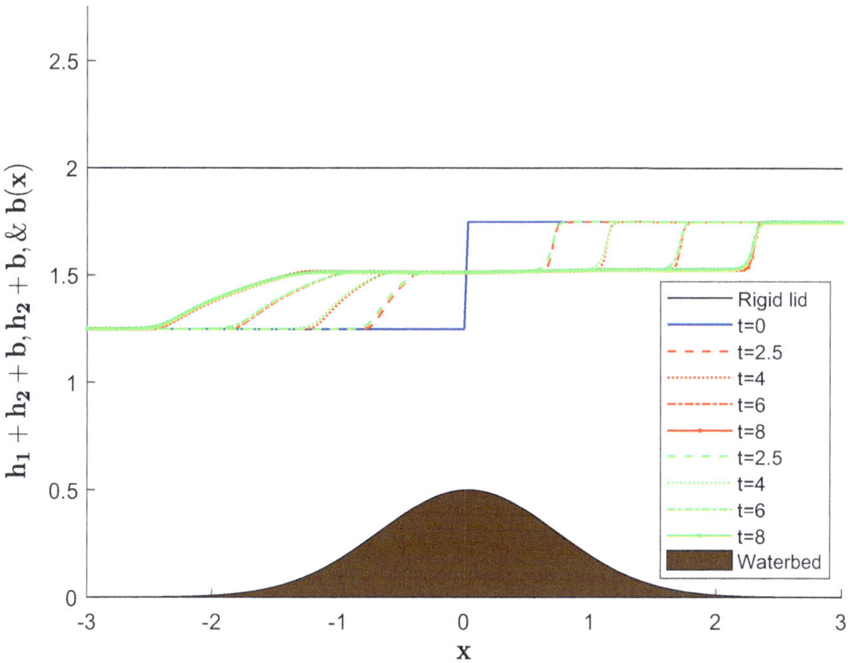

Fig. 12 Test case 4: Comparison between the solution computed using the UCS-SGM scheme (red color) and a MUSCL-RK2 scheme (green color) with 200 grid points at time $t \in \{2.5, 4, 6, 8\}$

Figure 13 shows the two-layer fluid elevation functions at the initial time (top) and at the final time (bottom) obtained with the aid of the UCS scheme. Oscillations are observed in the vicinity of the waterbed discontinuity and the rigid-lid constraint is violated. Figure 14 shows the results for the fluids' elevation functions obtained with the aid of the UCS-SGM scheme at the final time. Clearly we see that the rigid-lid constraint remains satisfied at the discrete level and the nonphysical oscillations arising in Fig. 13 (bottom) disappeared completely due to the careful application of the surface gradient method in the forth- and back-projections of the h_2 component. Thanks to the SGM-based projections, the ℓ_∞-norm of the error in the computed rigid lid remains within 10^{-16} when the computational domain is discretized using 200 grid points, thus confirming the potential of the proposed method to solve two-layer rigid-lid problems up to machine accuracy. In Fig. 15, we present a comparison between the solution computed using the UCS-SGM scheme and the MUSCL-RK2 scheme at time $t_f = 8$. A good agreement is observed between the two solutions. Furthermore, it is evident that the UCS-SGM scheme exhibits superior accuracy in capturing the shock wave compared to the MUSCL-RK2 scheme, delivering a sharper representation.

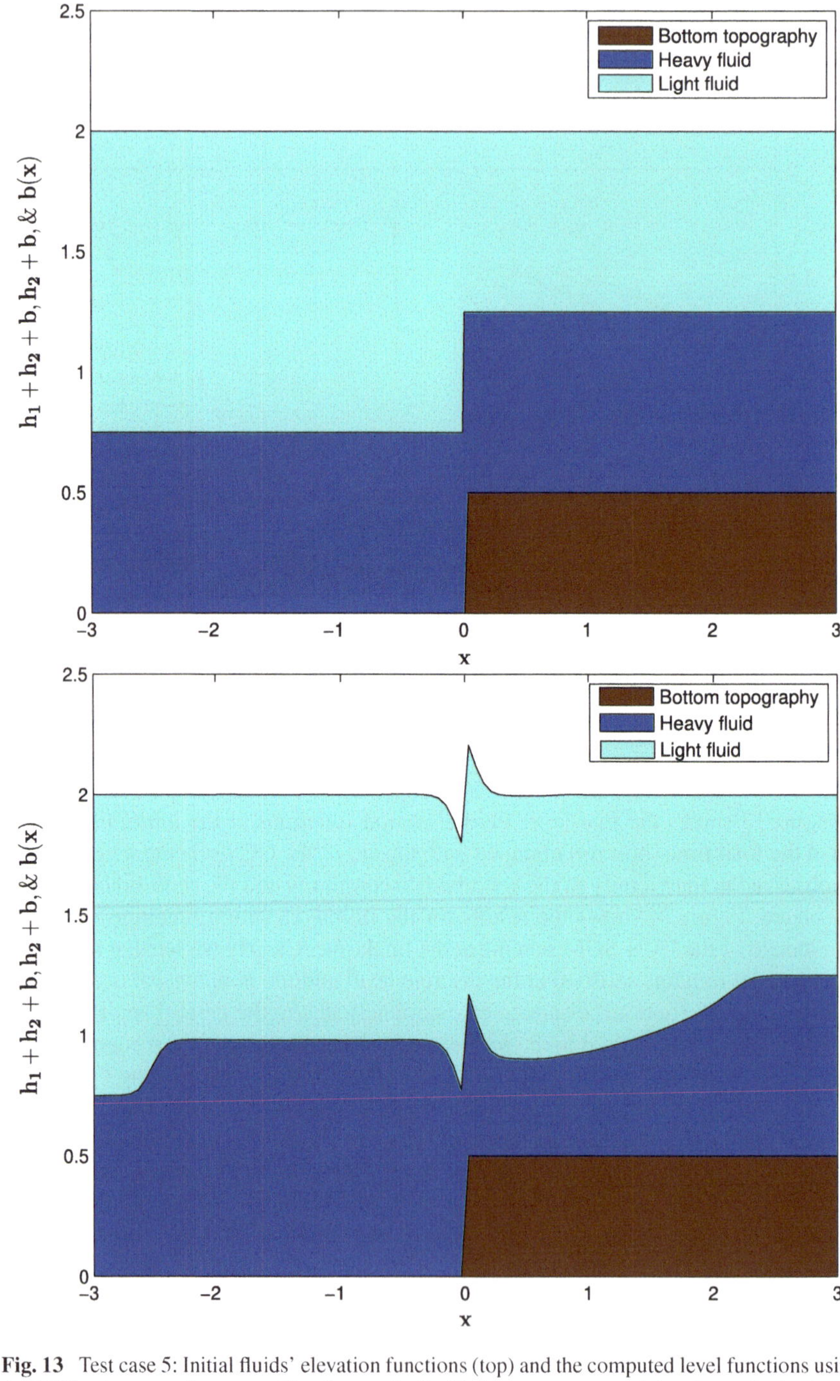

Fig. 13 Test case 5: Initial fluids' elevation functions (top) and the computed level functions using the UCS scheme at time $t_f = 10$ s (bottom). Oscillations occur around the discontinuity in the waterbed and the rigid-lid constraint is violated

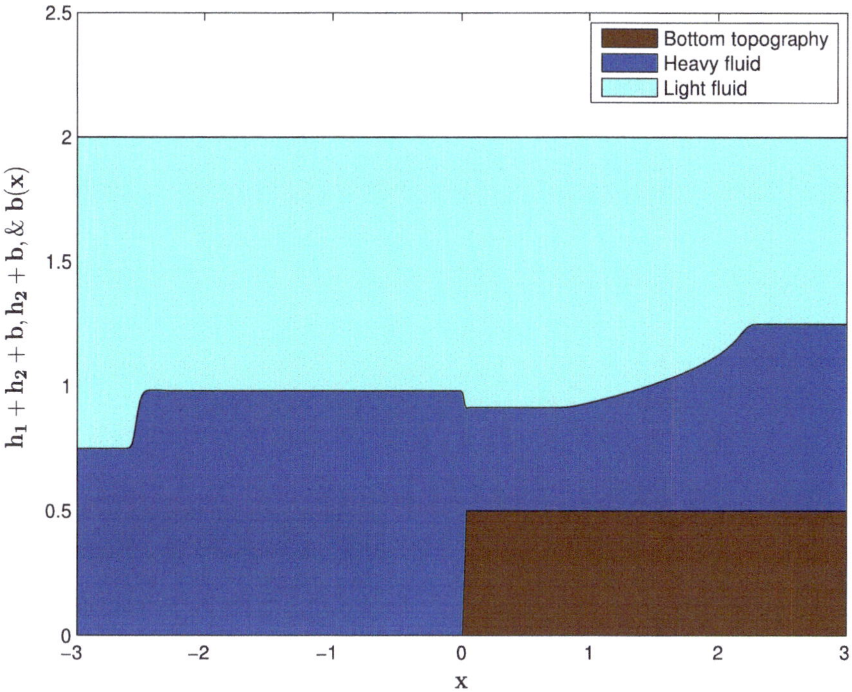

Fig. 14 Test case 5: Dam break problem set on a discontinuous waterbed. Profile of the two-layer fluid elevation functions obtained at time $t_f = 8$ s using the UCS-SGM scheme

5 Conclusion

This work is about a new formulation of the rigid-lid two-layer shallow water equations with variable topography. The proposed formulation is written as a system of hyperbolic conservation laws where the waterbed elevation function is embedded within the divergence of the flux term. We then extended our previously developed unstaggered central schemes for hyperbolic conservation laws to the case of two-layer rigid-lid systems and successfully solved some selected problems. The UCS scheme is not capable of effectively handling steady- or unsteady-state problems featuring either smooth or discontinuous waterbed functions. To this end, we extended the surface gradient method to the case of UCS schemes and successfully solved two-layer rigid-lid systems on either smooth or discontinuous topographies. We showed that the proposed numerical method is capable of handling, up to discrete level, some equilibrium solutions of the two-layer rigid-lid system. The numerical scheme was then implemented and validated and classical two-layer flow problems were successfully solved; the obtained numerical results were analyzed and discussed. The observed numerical results endorse the fact that the proposed numerical scheme can

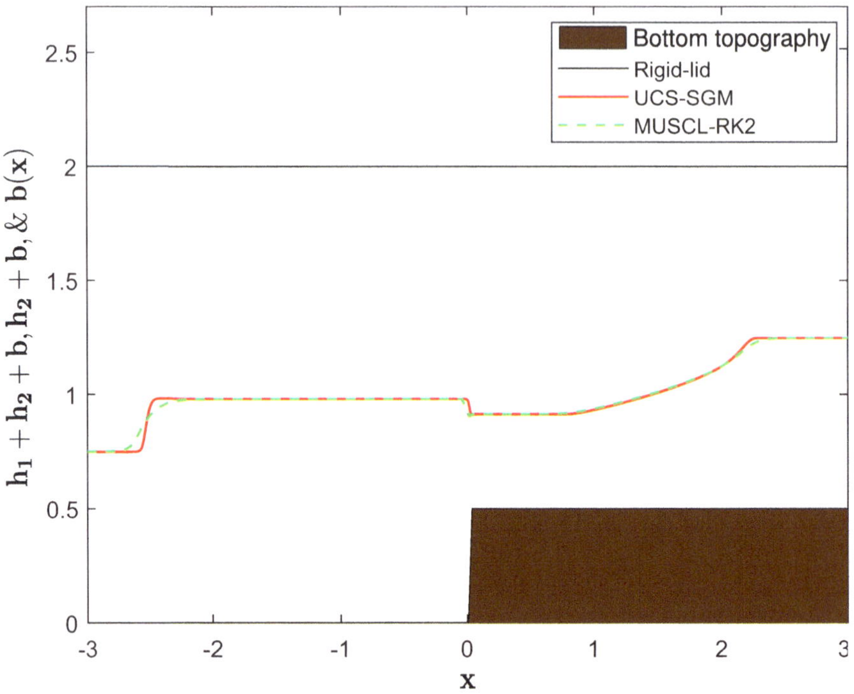

Fig. 15 Test case 5: Comparison between the solution computed using the UCS-SGM scheme and a MUSCL-RK2 scheme at time $t_f = 8$

effectively handle rigid-lid, two-layer flow problems on variable bathymetries both smooth and discontinuous.

References

1. R. Abgrall, S. Karni, Two-layer shallow water system: a relaxation approach. SIAM J. Sci. Comput. **31**, 1603–1627 (2009)
2. F. Alcrudo, P. Garcia-Navarro, A high-resolution godunov-type scheme in finite volumes for the 2d shallow-water equations. Int. J. Numer. Meth. Fluids **16**(6), 489–505 (1993)
3. C.T. Anh, Influence of surface tension and bottom topography on internal waves. Math. Models Methods Appl. Sci. **19**(12), 2145–2175 (2009). https://doi.org/10.1142/S0218202509004078
4. E. Audusse, F. Bouchut, M.O. Bristeau, R. Klein, B. Perthame, A fast and stable well-balanced scheme with hydrostatic reconstruction for shallow water flows. SIAM J. Sci. Comput. **25**(6), 2050–2065 (2004). https://doi.org/10.1137/S1064827503431090
5. T.B. Benjamin, Internal waves of permanent form in fluids of great depth. J. Fluid Mech. **29**(3), 559–592 (1967). https://doi.org/10.1017/S002211206700103X
6. A. Bermudez, M.E. Vazquez, Upwind methods for hyperbolic conservation laws with source terms. Comput. Fluids **23**(8), 1049–1071 (1994)

7. J. Bona, D. Lannes, J.C. Saut, Asymptotic models for internal waves. J. de Mathématiques Pures et Appliquées **89**(6), 538–566 (2008). https://doi.org/10.1016/j.matpur.2008.02.003. https://www.sciencedirect.com/science/article/pii/S0021782408000196
8. F. Bouchut, T. Morales de Luna, An entropy satisfying scheme for two-layer shallow water equations with uncoupled treatment. ESAIM: M2AN **42**(4), 683–698 (2008). https://doi.org/10.1051/m2an:2008019
9. C. Bourdarias, S. Gerbi, R. Lteif, A numerical scheme for an improved green-naghdi model in the camassa-holm regime for the propagation of internal waves. Comput. Fluids **156**, 283–304 (2017). https://doi.org/10.1016/j.compfluid.2017.07.012. Ninth international conference on computational fluid dynamics (ICCFD9)
10. A. Canestrelli, A. Siviglia, M. Dumbser, E.F. Toro, Well-balanced high-order centered schemes for non-conservative hyperbolic systems. Applications to shallow water equations with fixed and mobile bed. Adv. Water Resourc. **32**(6), 834–844 (2009)
11. M.J. Castro, J.A. Garcıa-Rodrıguez, J.M. González-Vida, J. Macıas, C. Parés, M.E. Vázquez-Cendón, Numerical simulation of two-layer shallow water flows through channels with irregular geometry. J. Comput. Phys. **195**(1), 202–235 (2004). https://doi.org/10.1016/j.jcp.2003.08.035
12. M.J. Castro-Díaz, E.D. Fernández-Nieto, J.M. González-Vida, C. Parés-Madroñal, Numerical treatment of the loss of hyperbolicity of the two-layer shallow-water system. J. Sci. Comput. **48**, 16–40 (2011). https://doi.org/10.1007/s10915-010-9427-5
13. M. Castro, J. Macías, C. Parés, A q-scheme for a class of systems of coupled conservation laws with source term. Application to a two-layer 1-d shallow water system. ESAIM: M2AN **35**(1), 107–127 (2001). https://doi.org/10.1051/m2an:2001108
14. W. Choi, R. Camassa, Weakly nonlinear internal waves in a two-fluid system. J. Fluid Mech. **313**, 83–103 (1996). https://doi.org/10.1017/S0022112096002133
15. W. Choi, R. Camassa, Fully nonlinear internal waves in a two-fluid system. J. Fluid Mech. **396**, 1–36 (1999). https://doi.org/10.1017/S0022112099005820
16. W. Craig, P. Guyenne, H. Kalisch, Hamiltonian long-wave expansions for free surfaces and interfaces. Commun. Pure Appl. Math. **58**(12), 1587–1641 (2005). https://doi.org/10.1002/cpa.20098
17. W. Craig, P. Guyenne, C. Sulem, Coupling between internal and surface waves. Nat. Hazards **57**, 617–642 (2011)
18. N. Črnjarić-Žic, S. Vuković, L. Sopta, Balanced central nt schemes for the shallow water equations, in *Proceedings of the Conference on Applied Mathematics and Scientific Computing* (Springer, 2005), pp. 171–185
19. V. Duchêne, Asymptotic shallow water models for internal waves in a two-fluid system with a free surface. SIAM J. Math. Anal. **42**(5), 2229–2260 (2010). https://doi.org/10.1137/090761100
20. V. Duchêne, Boussinesq/Boussinesq systems for internal waves with a free surface, and the KdV approximation. ESAIM. Math. Model. Numer. Anal. **46**(1), 145–185 (2012). https://doi.org/10.1051/m2an/2011037
21. V. Duchêne, Decoupled and unidirectional asymptotic models for the propagation of internal waves. Math. Models Methods Appl. Sci. **24**(1), 1–65 (2014). https://doi.org/10.1142/S0218202513500462
22. V. Duchêne, On the rigid-lid approximation for two shallow layers of immiscible fluids with small density contrast. J. Nonlinear Sci. **24**(4), 579–632 (2014). https://doi.org/10.1007/s00332-014-9200-2
23. V. Duchêne, S. Israwi, R. Talhouk, Shallow water asymptotic models for the propagation of internal waves. Discret. Continuous Dyn. Syst. Ser. S **7**(2), 239–269 (2014). https://doi.org/10.3934/dcdss.2014.7.239
24. V. Duchêne, S. Israwi, R. Talhouk, A new fully justified asymptotic model for the propagation of internal waves in the Camassa-Holm regime. SIAM J. Math. Anal. **47**(1), 240–290 (2015). https://doi.org/10.1137/130947064
25. M. Dumbser, V. Casulli, A staggered semi-implicit spectral discontinuous galerkin scheme for the shallow water equations. Appl. Math. Comput. **219**(15), 8057–8077 (2013)

26. A.E. Gill, Atmosphere-ocean dynamics, in *International Geophysics*, vol. 30 (Academic Press, 1982)
27. J.M. Greenberg, A.Y. Leroux, A well-balanced scheme for the numerical processing of source terms in hyperbolic equations. SIAM J. Numer. Anal. **33**(1), 1–16 (1996)
28. R. Grimshaw, E. Pelinovsky, O. Poloukhina, Higher-order korteweg-de vries models for internal solitary waves in a stratified shear flow with a free surface. Nonlinear Process. Geophys. **9**(3/4), 221–235 (2002). https://doi.org/10.5194/npg-9-221-2002
29. P. Guyenne, D. Lannes, J.C. Saut, Well-posedness of the Cauchy problem for models of large amplitude internal waves. Nonlinearity **23**(2), 237–275 (2010). https://doi.org/10.1088/0951-7715/23/2/003
30. K.R. Helfrich, W.K. Melville, Long nonlinear internal waves. Ann. Rev. Fluid Mech. **38**, 395–425. Annual Reviews, Palo Alto, CA (2006). https://doi.org/10.1146/annurev.fluid.38.050304.092129
31. G.H. Keulegan, Characteristics of internal solitary waves. J. Res. Natl. Bur. Stand. **51**(3), 133–140 (1953)
32. J. Kim, R.J. LeVeque, Two-layer shallow water system and its applications, hyperbolic problems: theory, numerics and applications, in *Proceedings of the twelth international conference on hyperbolic problems, Maryland*, vol. 67 (2009), pp. 737–743. https://doi.org/10.1090/psapm/067.2/2605269
33. C.G. Koop, G. Butler, An investigation of internal solitary waves in a two-fluid system. J. Fluid Mech. **112**, 225–251 (1981)
34. A. Kurganov, G. Petrova, Central-upwind schemes for two-layer shallow water equations. SIAM J. Sci. Comput. **31**(3), 1742–1773 (2009). https://doi.org/10.1137/080719091
35. D. Leonardi, *Internal and Surface Waves in a Two-Layer Fluid* (University of Illinois at Chicago, Thesis, 2011)
36. R. Lteif, Well-balanced numerical resolution of the two-layer shallow water equations under rigid-lid with wet-dry fronts. Comput.Fluids **235**, 105277 (2022). https://doi.org/10.1016/j.compfluid.2021.105277. https://www.sciencedirect.com/science/article/pii/S0045793021003704
37. R. Lteif, S. Israwi, R. Talhouk, An improved result for the full justification of asymptotic models for the propagation of internal waves. Commun. Pure Appl. Anal. **14**(6), 2203–2230 (2015). https://doi.org/10.3934/cpaa.2015.14.2203
38. X. Lu, B. Dong, B. Mao, X. Zhang, A robust and well-balanced numerical model for solving the two-layer shallow water equations over uneven topography. Comptes Rendus Mécanique **343**(7), 429–442 (2015). https://doi.org/10.1016/j.crme.2015.05.002. https://www.sciencedirect.com/science/article/pii/S163107211500056X
39. K.T. Mandli, A numerical method for the two layer shallow water equations with dry states. Ocean Model. **72**, 80–91 (2013). https://doi.org/10.1016/j.ocemod.2013.08.001. https://www.sciencedirect.com/science/article/pii/S1463500313001212
40. J.C. McWilliams, Modeling the oceanic general circulation. Annu. Rev. Fluid Mech. **28**(1), 215–248 (1996). https://doi.org/10.1146/annurev.fl.28.010196.001243
41. H. Michallet, E. Barthelemy, Experimental study of interfacial solitary waves. J. Fluid Mech. **366**, 159–177 (1998)
42. H. Nessyahu, E. Tadmor, Non-oscillatory central differencing for hyperbolic conservation laws. J. Comput. Phys. **87**(2), 408–463 (1990)
43. H.Y. Nguyen, F. Dias, A Boussinesq system for two-way propagation of interfacial waves. Phys. D **237**(18), 2365–2389 (2008). https://doi.org/10.1016/j.physd.2008.02.020
44. H. Ono, Algebraic solitary waves in stratified fluids. J. Phys. Soc. Jpn. **39**(4), 1082–1091 (1975). https://doi.org/10.1143/JPSJ.39.1082
45. L. Ovsyannikov, Two-layer shallow water model. J. Appl. Mech. Tech. Phys. **20**, 127–135 (1979). https://doi.org/10.1007/BF00910010
46. R. Touma, Central unstaggered finite volume schemes for hyperbolic systems: Applications to unsteady shallow water equations. Appl. Math. Comput. **213**(1), 47–59 (2009)

47. R. Touma, S. Khankan, Well-balanced unstaggered central schemes for one and two-dimensional shallow water equation systems. Appl. Math. Comput. **218**(10), 5948–5960 (2012)
48. R. Touma, C. Klingenberg, Well-balanced central finite volume methods for the ripa system. Appl. Numer. Math. **97**, 42–68 (2015)
49. R. Touma, U. Koley, C. Klingenberg, Well-balanced unstaggered central schemes for the euler equations with gravitation. SIAM J. Sci. Comput. **38**(5), B773–B807 (2016)
50. B. Van Leer, Towards the ultimate conservative difference scheme. IV. A new approach to numerical convection. J. Comput. Phys. **23**(3), 276–299 (1977)
51. Y. Xing, X. Zhang, C.W. Shu, Positivity-preserving high order well-balanced discontinuous galerkin methods for the shallow water equations. Adv. Water Resour. **33**(12), 1476–1493 (2010)
52. J.G. Zhou, D.M. Causon, C.G. Mingham, D.M. Ingram, The surface gradient method for the treatment of source terms in the shallow-water equations. J. Comput. Phys. **168**(1), 1–25 (2001)
53. F. Kanbar, R. Touma, C. Klingenberg, Well-balanced central scheme for the system of MHD equations with gravitational source term. Commun. Comput. Phys. **32**(3), 878–898 (2022)

Compressible Enstrophy Transport for Flow in a Low-Pressure Turbine with Unsteady Wakes Impinging at the Inflow

Aditi Sengupta

Abstract The current numerical research examines the role of wake amplitude on the underlying flow characteristics of the separation-induced transition and vorticity dynamics inside a T106A low-pressure turbine (LPT) blade passage. The study solves the two-dimensional (2D) compressible Navier-Stokes equations (NSE) using highly accurate compact schemes that satisfy the numerical dispersion relation. Four wake amplitudes, ranging from 0.20 to 0.80, are imposed at the inflow to explore the role of unsteady wakes on the flow separation on the suction surface. A periodic Gaussian wake, with a specified time period of wake passing and width of the Gaussian profile, is imposed at the inflow. The emergence and evolution of unsteady separation bubbles along the suction surface of the T106A blade are analyzed, revealing an increasing trend with wake amplitude. An overall increase in turbulent mixing is observed with increasing wake amplitude, leading to a thicker boundary layer on the suction surface. By applying the compressible enstrophy transport equation (CETE) to the T106A blade passage, the study explores the enstrophy dynamics in the flow. It reveals that the wake amplitude primarily affects the baroclinic vorticity and viscous terms of the CETE budget, with the highest contribution from the term due to viscous stresses, followed by baroclinic vorticity generation in the early stages of disturbance evolution. Other terms of the CETE budget show a nuanced dependency on the wake amplitude, paving the way for future investigations into the role of unsteady wakes on enstrophy dynamics.

1 Introduction

The phenomenon of transition is commonly observed in external aerodynamic and turbomachinery flows. It is found to significantly impact the operational performance of aircraft engines. In stages of turboengines operating at low pressures, this becomes particularly true as the flow experiences moderate Reynolds numbers (Re)

A. Sengupta (✉)
Department of Mechanical Engineering, Indian Institute of Technology Dhanbad, Dhanbad, India
e-mail: aditi@iitism.ac.in

D. Zeidan et al. (eds.), *Computational Fluid Dynamics: Novel Numerical and Computational Approaches*, Infosys Science Foundation Series,
https://doi.org/10.1007/978-981-97-8152-2_3

during acceleration for which the flow remains transitional over a large extent of a low-pressure turbine (LPT) blade. The overall engine efficiency has an overarching contribution from the LPT efficiency, so much so that a 0.7% increment in efficiency of the engine can be traced back to a 1% improvement in LPT's efficiency [1]. Current engine design trends involve LPT blades being subjected to higher loading per blade compared to designs of the past. This has been done with the aim of increasing fuel efficiency by bringing down the weight of the engine. However, this has the unfortunate tendency of making the blade more sensitive to flow separation, particularly on the suction surface. Focusing on the suction surface in particular, the total losses experienced have a 60% contribution from the boundary layer-related losses [2]. These boundary layer events are greatly influenced by the various environments external to the LPT blade. Thus, designing more effective LPT systems involves a careful consideration of the flow physics unraveled in the presence of adverse pressure gradients, which is typical on the suction surface of the blade. The transition process, in particular, has received attention from engine designers.

Initially, blade surfaces are hydrodynamically smooth upon deployment but these show a propensity to impart roughness due to various mechanisms by which the surface is damaged [3]. Bons [4] found that for the LPT blade, the suction surface shows a higher impact on the blade's performance due to progressive roughness [4]. Interestingly, for certain flow scenarios (such as low Re), the roughness can in fact prevent flow separation on the suction surface by triggering early transition and early reattachment. The roughness-induced transition is capable of suppressing laminar boundary layer separation which aids in lowering the overall blade losses, compared to the smooth blade surface. This benefit observed for low Re is dependent on the loading the blade is subjected to - with LPTs (more often than not) demonstrating this reduction in blade loss. Surface roughness is not the sole contributor to this suppression of the separated region [5], the same can be seen in the presence of free-stream excitation [6–8], unsteady wakes [9–11], and aeroelastic flutter phenomena [7, 10]. The current study investigates whether such a benefit can be achieved in the presence of unsteady wakes at the inflow, which will have implications for lowered profile loss for an LPT. The bypass transition mechanism, which governs the LPT flow in the presence of free-stream excitation [12, 13] and roughness, has received a thorough deliberation showing that a nonmodal route is often advocated, characterized by a series of unsteady separation bubbles [15]. However, the transition route with unsteady wakes at the inflow, particularly as a function of wake amplitude, has not been previously studied and will be a focus of the present chapter. Most studies focus on time-averaged boundary layer characteristics to explore the role of unsteady wakes at the inflow. Here, we will use a compressible enstrophy equation (CETE) [16] applied directly to the instantaneous flow field to explore the role of wake amplitude on the individual budget terms making up the equation.

A surmountable amount of the research pertaining to separation-induced transition in LPTs can be traced back to those involving idealized LPT blade configurations comprising of a flat plate upon which a streamwise pressure gradient is imposed, reminiscent of the pressure distribution on the suction surface [5, 6, 8]. Contrary to this, the T106A blade is a "first-generation" high lift blade based on the midsection of

Pratt and Whitney's PW2037, and thus represents an actual LPT blade. This offers a gradual loading on the suction surface, compared to the flat plate idealization which mimics an ultra-high lift configuration. The T106A blade has been studied experimentally [17] and computationally [9, 13] in the past. For a moderate *Re* of 51831, a numerical study of flow in a T106A LPT passage [19] demonstrated a complex fluid dynamical scenario triggered by the pressure distribution on the suction side. Multiple unsteady separation events were captured evoking numerous cycles of transition from laminar to turbulent flow. Rotor and stator interactions and the effect these may have on the flow field inside a T106A passage were studied by direct numerical simulations [20]. With the aid of twelve simulations, the effect of free-stream excitation on a T106A LPT cascade has been analyzed in [21] by varying turbulence levels and frequencies of the periodic unsteady wakes impinging at the inflow, and the Re. Interestingly, some numerical studies determine whether the Boussinesq assumption is valid for this LPT flow, showing that it is not the case near the trailing edge. A higher magnitude of losses is observed when the gaps between stator and rotors in a T106A cascade are smaller. This is attributed to the amplification of the turbulent kinetic energy for the wakes impinging at the inflow along the blade passage [22]. The linear LPT T106A cascade with surface roughness reveals via large eddy simulations that the peaks of the roughness profiles trigger uneven locations for the flow transition along the span [23]. A further analysis of the wake-induced transition was via a proper orthogonal decomposition which demonstrated that the losses on the suction side of the LPT blade originated from the finer scales [24]. For a chord Re of 30000, the T106A cascade with imposed free-stream turbulence at the inflow plane showed that the Klebanoff streaks were responsible for suppressing the separated flow, especially on the suction surface [25]. An analysis based on spectral proper orthogonal decomposition [26] showed that specific modes on the suction side's boundary layer are amplified via the propagation of upstream wakes. This also has the role of enhancing the turbulence downstream of the wake-induced transition.

A large body of literature pertaining to the suction surface boundary layer losses on LPT T106A blades is devoted to studying the effects of free-stream turbulence, surface roughness, unsteady wakes, and Reynolds numbers. However, a parametric investigation of the wake amplitude and its effect on the flow transition process (particularly on the suction side of the LPT blade) has not been done so far and the present research aims to achieve this. Four numerical implicit large eddy simulations of the flow inside a T106A blade passage are conducted by solving the two-dimensional (2D) compressible Navier-Stokes equations (NSE) for different amplitudes of a Gaussian wake, impinging at the inflow. One of the primary objectives of the current chapter will be to check how the amplitude of the Gaussian wake impacts the boundary layer parameters on the suction surface viz. the boundary layer thickness, unsteady separations, and the energy budget. This in turn will have potential implications for the blade's profile loss [27]. Does increasing wake amplitude delay or completely suppress separation on the suction surface? Does this impact the boundary layer losses, and if so, to what degree? These are the questions that are to be answered via the application of dispersion relation-preserving compact schemes as our numerical discretization method, widely adopted for their high accuracy in simulations. Additionally, we investigate the underlying mechanisms responsible for enstrophy generation and its distribution during various stages of the flow transition inside the T106A blade passage by applying the CETE [16]. This type of analysis has been in the past applied to external incompressible flows [18], for hydrodynamic instabilities [14] and in the lid-driven cavity [31]. Here, we intend to apply the CETE to an internal flow inside a turbine which is primarily driven by a streamwise pressure gradient, by adopting the T106A blade profile as our numerical setup.

2 Problem Formulation for Flow in a T106A Passage

The computational domain and boundary conditions utilized in this study are depicted in Fig. 1. These conditions are based on experiments conducted for a linear T106A cascade in a high-speed cascade wind tunnel at Universität der Bundeswehr, München, Germany, as documented in [17, 32]. The primary aim of these experiments was to provide benchmark data at low *Re* to validate numerical simulations. Because the T106A blade in this study experiences lighter loading compared to 'ultra-high lift blade designs, the boundary layer on the suction side remains relatively thin. A no-slip condition is enforced on the blade surface, and periodicity is imposed in the pitchwise direction. The computational domain's inlet is positioned at a distance equivalent to 0.4 times the axial chord length upstream of the leading

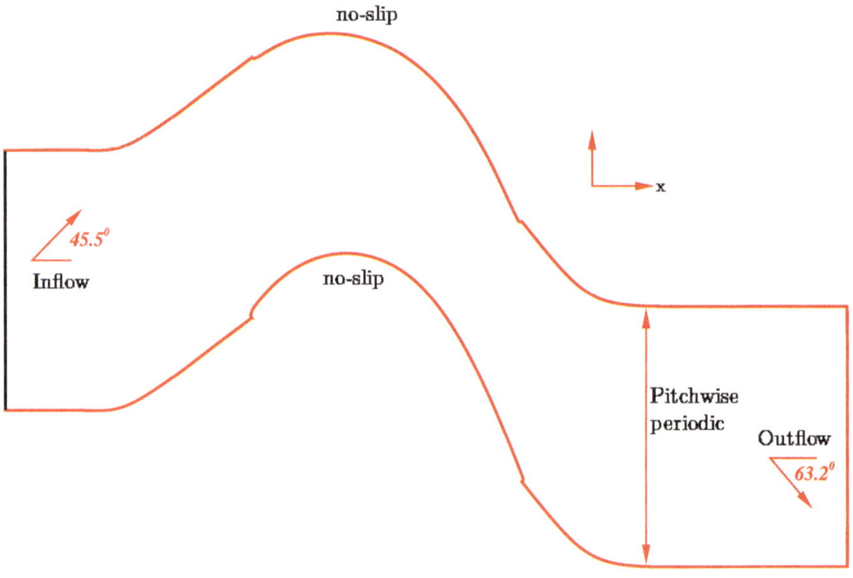

Fig. 1 Schematic of the computational domain and boundary conditions used to simulate flow in a T106A LPT cascade with unsteady wakes at the inflow

edge, while the outlet extends up to 0.5 times the axial chord length downstream of the trailing edge. To discretize the computational domain, an H-type structured grid is employed, comprising a total of 1251 × 501 grid points in the streamwise, wall-normal, and spanwise directions. Among these, the streamwise direction is divided into 276 points upstream of the leading edge, 751 points along the blade surface, and 226 points downstream of the blade. The mesh used here is based on a set of simulations calibrated with the experimental distributions of the time-averaged coefficient of pressure (C_p) on the suction and pressure surfaces of the T106A blade [18]. The validation efforts showed that the current grid is capable of capturing all essential flow physics to a reasonable degree of accuracy. The simulations are conducted at a Re of 1.6×10^5, which is based on the actual chord length (c) and the exit velocity (U_{TE}), following the methodology from Wissink's DNS study [9]. The blade is mid-loaded, with peak suction occurring at a position corresponding to 42% of the suction surface length (S_0). The pitch between blades is set at 0.9306 times the chord length (which is the length scale of the present set of simulations). At the inflow boundary, where the flow impinges with an angle of attack ($\alpha_{in} = 45.5°$), an inflow Mach number (M_s) is defined. This Mach number corresponds to Re of 60142.72. The outflow boundary is defined with an angle of attack ($\alpha_{out} = -63.2°$) and an outflow Mach number (M_{out}). Specifically, the inflow and outflow Mach numbers are set to 0.15 and 0.404, respectively. To prevent spurious reflections from contaminating the inflow and outflow boundaries, a characteristic-based boundary condition utilizing Riemann invariants, as described in [33], is implemented.

2.1 Governing Equations

We solve the unsteady 2D compressible NSE, which satisfies the conservation principles of mass, momentum, and energy. These fundamental equations are expressed in the strong conservation form. In this form, any symmetric discretization method has no impact on the solution within the interior, as neighboring cells seamlessly merge into each other, leaving only the boundary terms and transforming the scenario into a true boundary value problem. In the context of compressible flows, when discontinuities arise in the physical variables, the fluxes maintain continuity, while the physical variables themselves may exhibit discontinuities. Computing continuous functions is more straightforward compared to handling discontinuous functions, which have a broader bandwidth in wavenumber space. Importantly, this form remains unchanged even when the equations of motion are transformed into an arbitrary curvilinear coordinate system. Thus, the governing equations are articulated in the strong conservation form as follows:

$$\frac{\partial \hat{Q}}{\partial t^*} + \frac{\partial \hat{E}}{\partial x^*} + \frac{\partial \hat{F}}{\partial y^*} = \frac{\partial \hat{E}_v}{\partial x^*} + \frac{\partial \hat{F}_v}{\partial y^*} \tag{1}$$

where the conserved variables are given by

$$\hat{Q} = \left[\rho^*;\ \rho^* u^*;\ \rho^* v^*;\ \rho^* e_t^*\right]^T \tag{2}$$

The convective flux variables $\hat{E}$ and $\hat{F}$ are given as

$$\hat{E} = \left[\rho^* u^*;\ \rho^* u^{*2} + p^*;\ \rho^* u^* v^*;\ (\rho^* e_t^* + p^*) u^*\right]^T \tag{3}$$

$$\hat{F} = \left[\rho^* v^*;\ \rho^* u^* v^*;\ \rho^* v^{*2} + p^*;\ (\rho^* e_t^* + p^*) v^*\right]^T \tag{4}$$

and the viscous flux vectors $\hat{E}_v$ and $\hat{F}_v$ are given as

$$\hat{E}_v = \left[0;\ \tau_{xx}^*;\ \tau_{xy}^*;\ u^* \tau_{xx}^* + v^* \tau_{xy}^* - q_x^*\right]^T \tag{5}$$

$$\hat{F}_v = \left[0;\ \tau_{yx}^*;\ \tau_{yy}^*;\ u^* \tau_{yx}^* + v^* \tau_{yy}^* - q_y^*\right]^T \tag{6}$$

The variables ρ^*, p^*, u^*, v^*, T^*, and e_t^* correspond to the dimensional quantities of density, fluid pressure, Cartesian components of fluid velocity, absolute temperature, and the specific total energy of the fluid, respectively. These are measured in kg/m^3, N/m^2 or Pa, m/s, Kelvin and Joules, respectively. In the current formulation for air, the specific heat ratio is represented by γ and is set to 1.4. The components of the symmetric viscous stress tensor are denoted as τ_{xx}^*, τ_{yy}^*, τ_{xy}^*, and τ_{yx}^* (measured in N/m^2), and they are interrelated with the velocity gradients as follows:

$$\tau_{xx}^* = \left(2\mu \frac{\partial u^*}{\partial x^*} + \lambda \nabla^* \cdot \mathbf{V}^*\right),\ \tau_{yy}^* = \left(2\mu \frac{\partial v^*}{\partial y^*} + \lambda \nabla^* \cdot \mathbf{V}^*\right), \tag{7}$$

$$\tau_{xy}^* = \tau_{yx}^* = \mu \left(\frac{\partial u^*}{\partial y^*} + \frac{\partial v^*}{\partial x^*}\right) \tag{8}$$

In this formulation, fluid properties such as specific heat capacity (C_v) and thermal conductivity (κ) are treated as constants. The expressions for the heat conduction terms q_x^* and q_y^* (expressed in Watts per meter square W/m^2) are as follows:

$$q_x^* = -\kappa \frac{\partial T^*}{\partial x^*},\ q_y^* = -\kappa \frac{\partial T^*}{\partial y^*} \tag{9}$$

Stokes' hypothesis establishes a connection between molecular viscosity, denoted as μ (measured in $Pa - s$), and the second coefficient of viscosity, represented as λ, through the relationship: $\lambda = -2\mu/3$. This relationship is commonly employed as the constitutive relation. To determine the viscosity as a function of temperature, Sutherland's law is applied. The closure for the system of equations is achieved through the ideal gas equation, as the working fluid is air for the simulations. If one were to operate with any other working fluid, we would need to additionally solve a pressure Poisson equation for solving the pressure and velocity simultaneously. The ideal gas law, serving as the equation of state here, is given by

$$p^* = \rho^* R^* T^* \tag{10}$$

and it is used to define specific energy, e_t^* as

$$e_t^* = \frac{p^*}{[\rho^*(\gamma - 1)]} + \frac{(u^{*2} + v^{*2})}{2} \tag{11}$$

The temperature and density scales used for non-dimensionalization are $T_s = 300$ K and $\rho_s = 1.177\,\mathrm{kg/m^3}$. These are chosen based on typical temperatures and densities measured in experiments for an LPT cascade at the inflow [37], where the working fluid is air.

2.2 *Imposing Gaussian Wakes at the Inflow*

In the experimental studies conducted by Lou and Hourmouziadis [34], unsteadiness was introduced by applying a sinusoidal perturbation to the free-stream flow. In our current investigation of wake-induced transition, we will adopt a similar approach, utilizing a periodic Gaussian function to introduce unsteadiness into the free-stream flow. This methodology aligns with the procedures employed in previous studies, as outlined in references [10, 35].

In the wake-induced transition simulations, we introduce a Gaussian wake at the inflow using the following velocity profile. This profile is added to the streamwise inlet velocity during each iteration.

$$u(w) = Ae^{-\alpha\left[mod\left(\frac{y}{y_{max}}+\frac{2t}{T_w}-1,2\right)\right]^2} \tag{12}$$

In this context, A represents the amplitude related to the wake, α is the exponential constant governing the width of the Gaussian profile, y_{max} signifies the maximum wall-normal height within the domain, and T denotes the time period for the passage of the wake.

A wake is a periodic disturbance characterized by high turbulence flow, creating a velocity deficit downstream of the blade. The transition process becomes intricate due to the inherent unsteadiness of multi-stage turbomachines. Wakes generated by one blade row convect through downstream blade passages, periodically disrupting the boundary layer. This cyclic passage of turbulent wakes has implications for the separation bubble observed on modern LPT blades. Under specific flow conditions, the wake passage leads to a reduction in profile loss [36]. The presence of turbulent wakes results in an overall restrained separation region. This can be attributed to turbulent spots induced by wakes upstream of the separation point, acting as a preventive measure against boundary layer separation. Additionally, the calmed region that ensues after the wake passage contributes to separation suppression by promoting elevated shear and complete velocity profiles. Investigating the detailed interaction between turbulent wakes and the unsteady separation bubble, which re-establishes between wake passing events, is crucial. Within the allowable wake passing frequencies for a row of upstream blade-induced wakes [37], we focus on studying the impact of wake amplitudes on separation-induced transitions in LPTs.

We utilize Eq. (12) to ensure that during each cycle of wake passage, the incoming wake takes into account the relative position of the LPT blade in relation to the stator blades, along with the associated phase shift of the wake, as described in [36]. For our simulations, we set the non-dimensional time period of wake passage to $T = 0.35$, and the exponential constant of the Gaussian function to $\alpha = 19$. These parameter values are chosen to replicate the wake spacing considered by Karaca and Gungor [35], while remaining within the acceptable limits of the non-dimensional Strouhal number for upstream blade-induced wakes. The non-dimensional Strouhal number associated with the wake is defined as $St = L_s/(U_s T)$, where L_s and U_s represent the length and velocity scales, respectively. In our case, these are the chord length (c) and the velocity at the trailing edge (U_{TE}), rendering $St = 0.34033$. The typical value observed for a wake originating from upstream blades is approximately 0.3 [37]. To investigate the influence of varying wake amplitudes on the separation-induced transition along the suction surface of a T106A cascade, we conducted four numerical simulations. These simulations are designed with different amplitudes for the Gaussian wake, and here are the specifics for each test case:

(i) Case-1: In this case, we introduced a Gaussian wake at the inflow with a non-dimensional amplitude of $A = 0.2$, time period of wake passing is 0.35, and width of the Gaussian is given by the exponential constant in Eq. (12) as $\alpha = 19$.

(ii) Case-2: This simulation involved doubling the wake amplitude compared to Case-1. The parameters are as follows for the Gaussian wake: non-dimensional amplitude of $A = 0.4$, time period of wake passing is 0.35, and width of the Gaussian is given by the exponential constant in Eq. (12) as $\alpha = 19$.

(iii) Case-3: For this case, we set the Gaussian wake's amplitude to three times that of Case-1. The parameters are as follows for the Gaussian wake: non-dimensional amplitude of $A = 0.6$, time period of wake passing is 0.35, and width of the Gaussian is given by the exponential constant in Eq. (12) as $\alpha = 19$.

(iv) Case-4: In this instance, we applied a Gaussian wake at the inflow with an amplitude that is four times that of Case-1. The parameters are as follows for the Gaussian wake: non-dimensional amplitude of $A = 0.8$, time period of wake passing is 0.35, and width of the Gaussian is given by the exponential constant in Eq. (12) as $\alpha = 19$.

The amplitude of the Gaussian wake can take values between 0 and 1. To describe a trend of flow physics as a function of wake amplitude, we have considered four representative cases across the permitted range. Each of these test cases was executed using 80 computational cores over a duration of 184 compute hours, resulting in a total of 14,720 core hours per test case. To ensure the removal of initial transients in the flow, we conducted the simulations for five to six complete through-flow cycles. Furthermore, for the purpose of time-averaging the flow field, we performed an additional 10 to 12 complete through-flow simulations and stored time-resolved data.

3 Numerical Methodology

The current simulations employ a compact scheme for spatial discretization, known for its dispersion relation preservation capabilities [28]. Specifically, the OUCS3 scheme (a second-order optimized upwind compact scheme) is utilized to calculate convective derivatives while maintaining equality between the numerical and physical dispersion relations across a broad range of wavenumbers and frequencies. This feature is particularly critical for wave-dominated flows, such as those observed on the suction surface of the LPT blade [7].

The upwinded interior stencil for the OUCS3 scheme is as follows:

$$m_{j-1}u'_{j-1} + u'_j + m_{j+1}u'_{j+1} = \frac{1}{h}\sum_{k=-2}^{2} n_k u_{j+k}$$

where u'_j is the spatial derivative evaluated at j^{th} node, $m_{j\pm1} = D \pm \frac{\eta}{60}$; $n_{\pm2} = \pm\frac{F}{4} + \frac{\eta}{300}$; $n_{\pm1} = \pm\frac{E}{2} + \frac{\eta}{30}$; and $n_0 = -\frac{11\eta}{150}$ with $D = 0.3793894912$. $E = 1.57557379$; $F = 0.183205192$ and $\eta = -2$. These parameters have been obtained by analyzing the numerical stability of the compact scheme used for spatial discretization and the time integration method by global spectral analysis [28]. This method helped to optimize the coefficients of the OUCS3 scheme by minimizing the different numerical errors with respect to a one-dimensional convection equation describing wave phenomena. This model wave equation is a non-dispersive, non-dissipative equation with an exact analytical solution, making it an effective benchmarking tool for numerical methods. It provides a dispersion relation between the spatial and temporal scales of a hyperbolic partial differential equation and is thus an effective tool for characterizing the nature of the Navier-Stokes equation being solved here.

To ensure accurate results in capturing the series of unsteady separation bubbles on the suction surface, a one-dimensional fifth-order filter with a filter coefficient of 0.47 is applied in both the x and y directions. For viscous flux calculations, the second-order central difference (CD_2) scheme is employed, and time integration is carried out using the classical four-stage, fourth-order Runge-Kutta method. The choice of time-step for these simulations is carefully selected to maintain the neutral stability [28] of the numerical methods, ensuring an accurate representation of the flow dynamics.

In the four-stage, fourth order, explicit Runge-Kutta method, the autonomous system, $\frac{\partial u}{\partial t} = f(u)$, is integrated from t^n to t^{n+1} using four increment functions, as follows:

$$K_1 = hf(u^n)$$
$$K_2 = hf(u^n + a_{21}K_1)$$
$$K_3 = hf(u^n + a_{31}K_1 + a_{32}K_2)$$
$$K_4 = hf(u^n + a_{41}K_1 + a_{42}K_2 + a_{43}K_3)$$

such that the time-marching yields

$$u^{n+1} = u^n + W_1K_1 + W_2K_2 + W_3K_3 + W_4K_4$$

where the parameters are evaluated using Taylor series expansion as follows:

$$a_{21} = \frac{1}{2}, a_{31} = 0, a_{32} = \frac{1}{2}, a_{41} = a_{42} = 0, a_{43} = 1.$$

$$W_1 = \frac{1}{6}, W_2 = \frac{1}{3}, W_3 = \frac{1}{3}, W_4 = \frac{1}{6}.$$

The accuracy of any discretization scheme can be better gauged by inspecting the compact scheme in wavenumber space [28]. If the function is represented as $f(x) = \int F(k)e^{ikx}dk$, then the exact representation of first derivative is given by

$$f'_{ex} = \int ikF(k)e^{ikx}dk$$

The discretized (numerical) derivative is represented in a similar notational form as

$$f'_{num} = \int ik_{eq}F(k)e^{ikx}dk$$

To assess the accuracy of spatial discretization schemes, one can examine the ratio of k_{eq} to k plotted against the non-dimensional wavenumber (kh), as detailed in [30]. Ideally, for representing discretization in the first derivative, the quotient k_{eq}/k should closely approach 1. Figure 2a illustrates this ratio along the ordinate, with the non-dimensional wavenumber kh extending up to the Nyquist limit of π along the abscissa. Notably, in Fig. 2a, the OUCS3 compact scheme proves to be the most accurate, outperforming several higher-order explicit central difference schemes. In this context, the upwinding of the OUCS3 scheme is disabled by setting

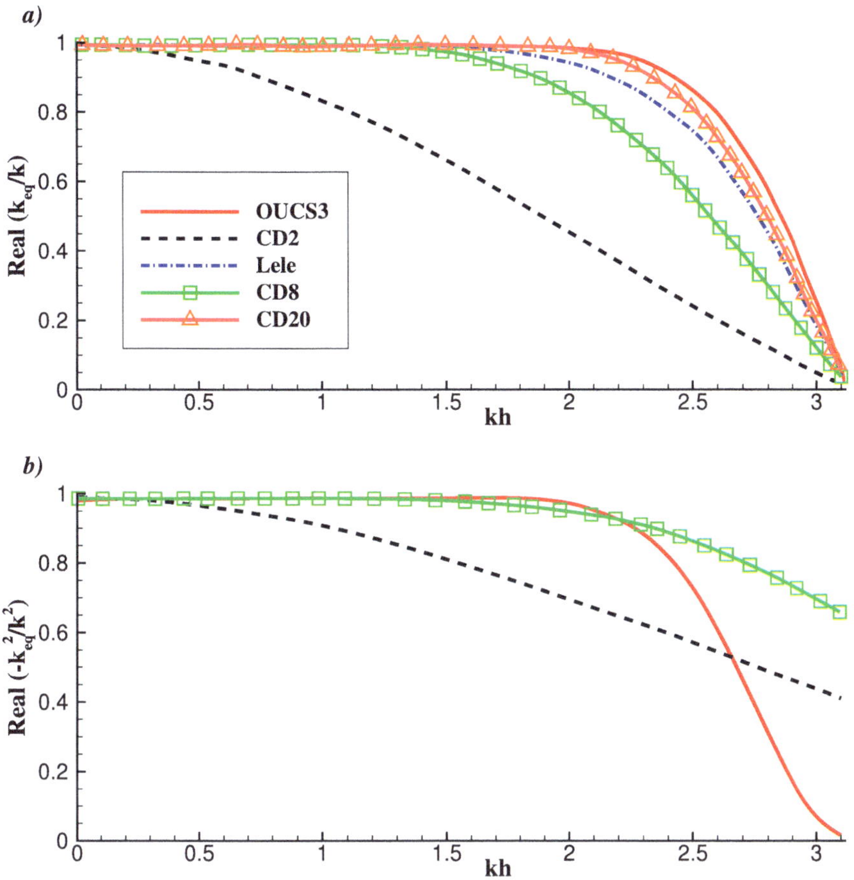

Fig. 2 The spectral accuracy of the indicated central explicit schemes are compared with the OUCS3 and sixth-order Lele's compact schemes by plotting k_{eq}/k as a function of the non-dimensional wavenumber (kh) for the first derivative in a frame (**a**). The accuracy in computing the second derivative by use of the indicated methods is shown in a frame (**b**)

$\eta = 0$, allowing the evaluation of the first derivative through a central scheme. It is also worth highlighting that the 20th-order central difference CD20 scheme surpasses Lele's sixth-order accurate compact scheme [29].

A similar comparison among different methods for the second derivative is depicted in Fig. 2b, where the real part of the ratio, $-k^{(2)}eq/k^2$, is plotted against the wavenumber kh. Here, the second derivative of any function is expressed as, $f'' = \int k^{(2)}eq F(k)e^{ikx}dk$. The central explicit schemes exhibit improved performance as the order of the methods increases. The OUCS3 scheme, applied twice to obtain the second derivative, exhibits inherent zero effectiveness at the Nyquist limit $(kh = \pi)$, as indicated in Fig. 2b. Furthermore, the OUCS3 and CD8 schemes demonstrate neutral stability (real part of $-k_{eq}^{(2)}/k^2$ equals one) over a more extensive range of wavenumbers, up to $kh = 2.4$ and 1.57, respectively, compared to the CD2 method. The OUCS3 scheme showcases high spectral accuracy, as the effectiveness of the numerical scheme remains equal to the exact derivative for lower wavenumbers, slightly exceeding the physical diffusion until $kh = 2.4$. This performance is notably superior to CD2 or CD8.

4 Vorticity Dynamics of Flow Inside a T106A Passage with Unsteady Wakes at the Inflow

In this section, we delve into the detailed analysis of the vorticity dynamics on the suction and pressure surfaces of the T106A blade. This examination focuses on how these dynamics vary with different wake amplitudes, denoted as A. This includes observing the formation and evolution of vortical structures, identifying regions of high vorticity, and assessing how changes in wake amplitude influence the overall flow field. Through this analysis, we seek to uncover the mechanisms by which wake disturbances affect the aerodynamic performance and stability of the turbine blade, providing insights that are crucial for optimizing blade design and enhancing turbine efficiency.

Figure 3 compares vorticity magnitude contours across four simulated test cases, highlighting the influence of wake amplitude on the flow dynamics. The selected time instant ($t = 40$) captures the flow behavior at an advanced stage of instability development. In all cases, a pre-transitional boundary layer forms from the leading edge, which has a minimal thickness on the suction side of the T106A blade. For Case-1, with an amplitude of 0.2, flow separation occurs beyond $S/S_0 = 0.42$, coinciding with the location on the suction surface which displays maximum suction, for the T106A blade. Near the trailing edge, small separation bubbles are detected, one attached to the boundary layer and another emerging beyond the trailing edge. Previous research [21] has also reported two unsteady separation bubbles in the proximity of the trailing edge on the suction side, under uniform inflow conditions. As the wake amplitude increases, a thicker boundary layer develops on the suction surface, accentuated by a series of unsteady separations along the suction surface. These separation events increase in both magnitude and number with greater wake amplitude, as documented in [38]. This sequence of unsteady separation bubbles is attributed to vortex-induced instability, previously reported for flows subjected to free-stream excitation [12, 15]. The wall-normal extents of these separation bubbles are significantly larger compared to those at lower amplitudes, with the bubbles breaking and merging over time. With higher Gaussian wake amplitudes at the inflow, a more pronounced free-stream disturbance is evident at the inlet. Additionally, the shedding of vortices from the pressure surface decreases in magnitude, with a higher number of vortical eddies and increased unsteadiness associated with larger wake amplitudes. This reduction in vortex shedding is reflected in lower maximum vorticity magnitude values as the wake amplitude increases.

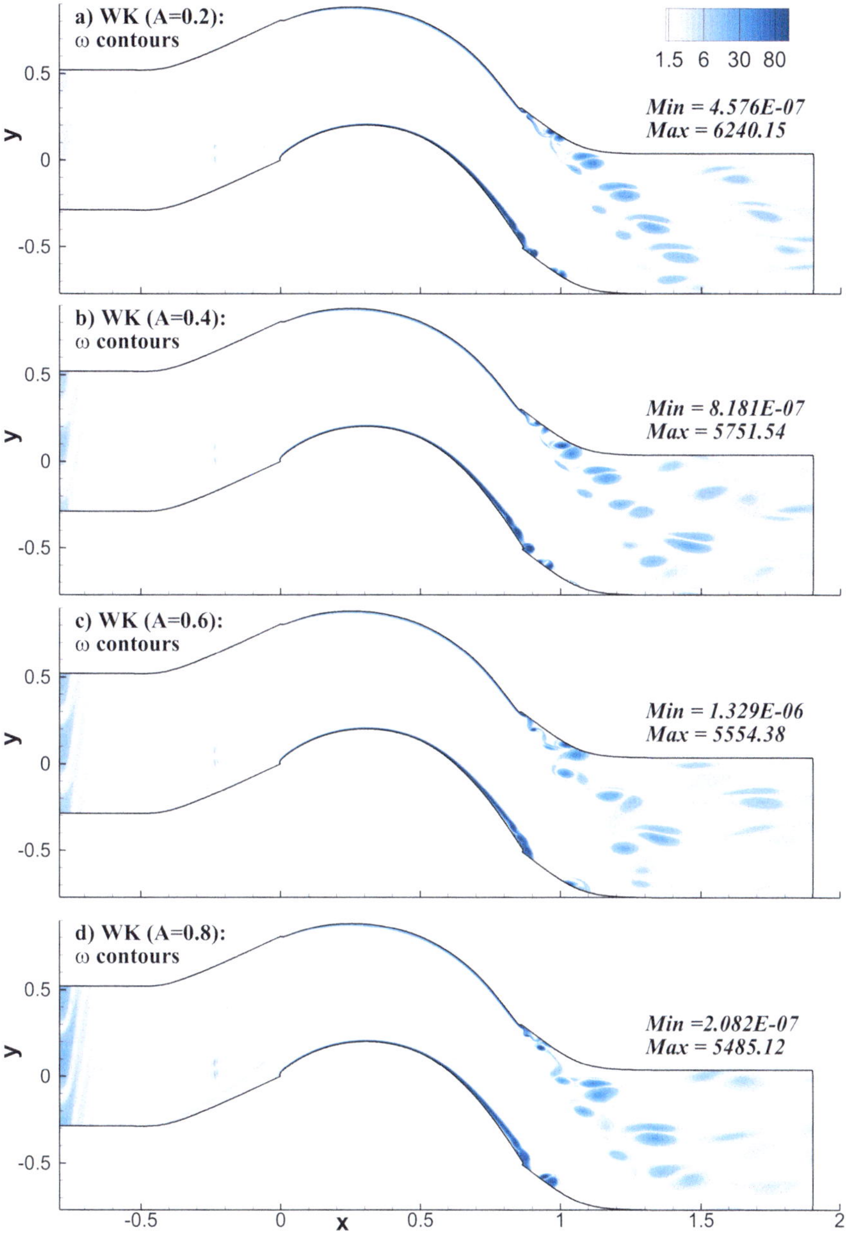

Fig. 3 Comparison of vorticity magnitude contours for cases with unsteady wakes impinging at the inflow with wake amplitude: **a** 0.2, **b** 0.4, **c** 0.6 and **d** 0.8

Figure 4 compares time series and spectra of spanwise vorticity on the suction surface of the T106A blade for the different wake amplitudes, $A = 0.2, 0.4, 0.6$, and 0.8. In Fig. 4a, for the case with $A = 0.2$, the time series reveals a dominant underlying time period along with unsteady fluctuations. This is corroborated by the spectrum in Fig. 4b, which displays distinct "peaks" within the frequency range of 14–16 Hz. These peaks are characteristic of Kelvin-Helmholtz vortex shedding [39], typical in flows with low levels of free-stream excitation. The amplitudes at lower frequencies are significantly smaller, about eight times less than the dominant frequency. As the wake amplitude increases to 0.4 in Fig. 4c, the time series shows multiple time periods, indicating a multi-periodic, complex, and chaotic dynamical behavior. This corroborates with the formation of multiple unsteady separation bubbles, triggered by the separation-induced transition on the suction surface [40]. The complexity is reflected in the spectrum shown in Fig. 4d, where vorticity is spread over a wide range of frequencies, in addition to the dominant peak. With a further increase in wake amplitude to 0.6 in Fig. 4e, the time series exhibits even greater unsteadiness, characterized by its multi-periodic and chaotic nature. The corresponding spectrum in Fig. 4f shows three significant peaks, with vorticity distributed over a larger number of frequencies compared to lower wake amplitudes. This suggests that as the level of free-stream excitation increases via increased wake amplitude, the flow transitions to turbulence earlier, with elevated unsteadiness and more unsteady separation events on the suction surface. Each coherent structure introduces its own time period, contributing to the multi-periodic nature of the spectrum. At even higher wake amplitudes, shown in Fig. 4g and h, the peak at lower frequencies becomes more pronounced. The wake passing frequency, 2.857 Hz, becomes evident in the spectrum as the wake amplitude increases. In Fig. 4h, the wake passing frequency dominates, along with the vortex shedding frequency of the Kelvin-Helmholtz eddies. This behavior demonstrates that the flow transition follows a nonmodal route of bypass transition, which has been previously documented in flows dominated by pressure gradients [15].

Figure 5 compares time series and spectra of spanwise vorticity on the pressure surface of the T106A blade for the different wake amplitudes, $A = 0.2, 0.4, 0.6$, and 0.8. As previously discussed with reference to Fig. 3, flow separation does not occur on the pressure surface of the T106A LPT blade. Due to the streamlined curvature on the pressure side, vortex shedding occurs. As a consequence of centrifugal forces, the flow accelerates on the convex pressure surface, which in turn reduces the turbulence inside the boundary layer. On the other hand, the curvature of the concave suction surface aids in the production of turbulence inside the boundary layer [19]. The time series data is collected from a region where vortex shedding originates from the trailing edge. For $A = 0.2$, the time series in Fig. 5a display high unsteadiness which can be traced back to the time periods of vortex shedding. Each vortex is linked to a specific time period, as can be seen in Fig. 5b, where the spectrum shows multiple peaks in the frequency plane. These are found to be sub-harmonics of the Kelvin-Helmholtz vortex shedding frequency range (14–16 Hz) [39]. The vorticity spectrum amplitude is higher on the suction side compared to the pressure surface, indicating greater susceptibility to boundary layer separation on the suction side in the presence of unsteady wakes. When the wake amplitude increases to 0.4, as

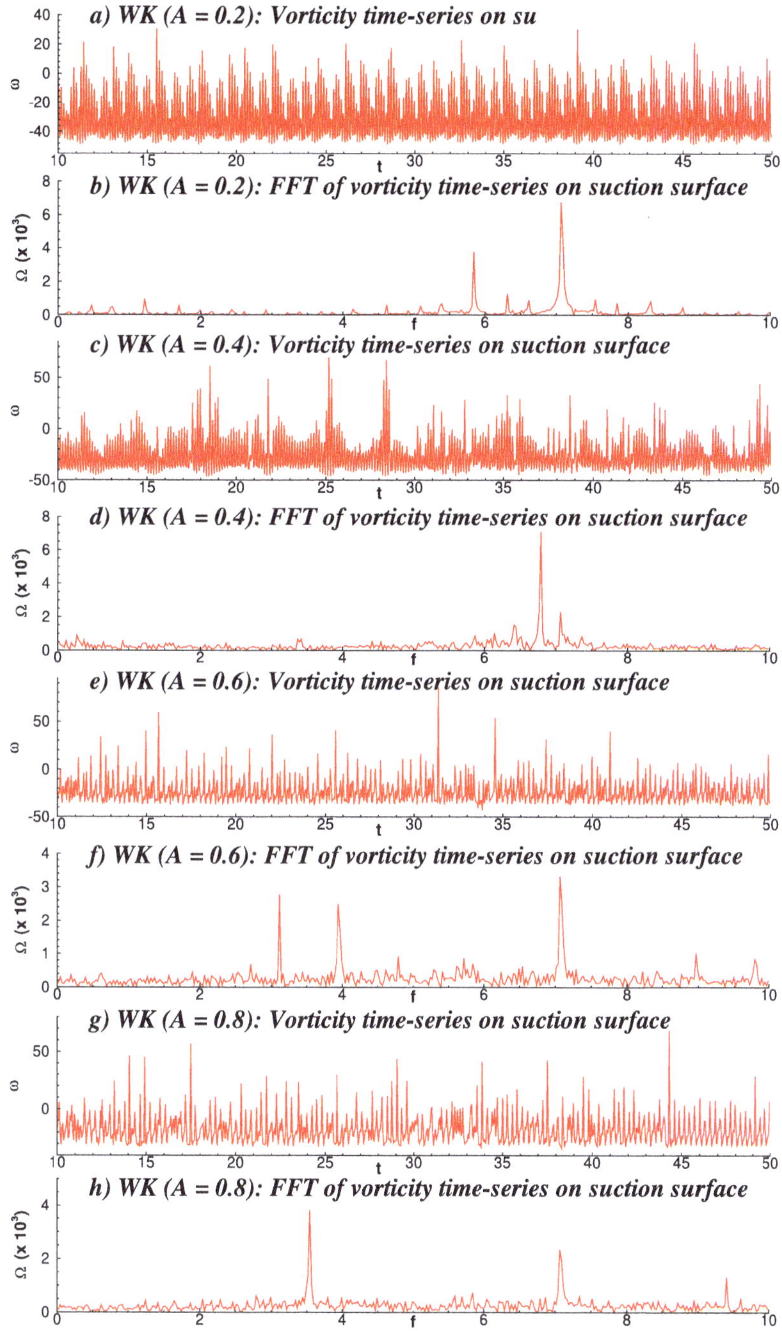

Fig. 4 Time series of vorticity (a, c, e, g) and its Fast Fourier transform (FFT) (b, d, f, h) on suction surface of T106A for cases with wake amplitude: 0.2, 0.4, 0.6 and 0.8

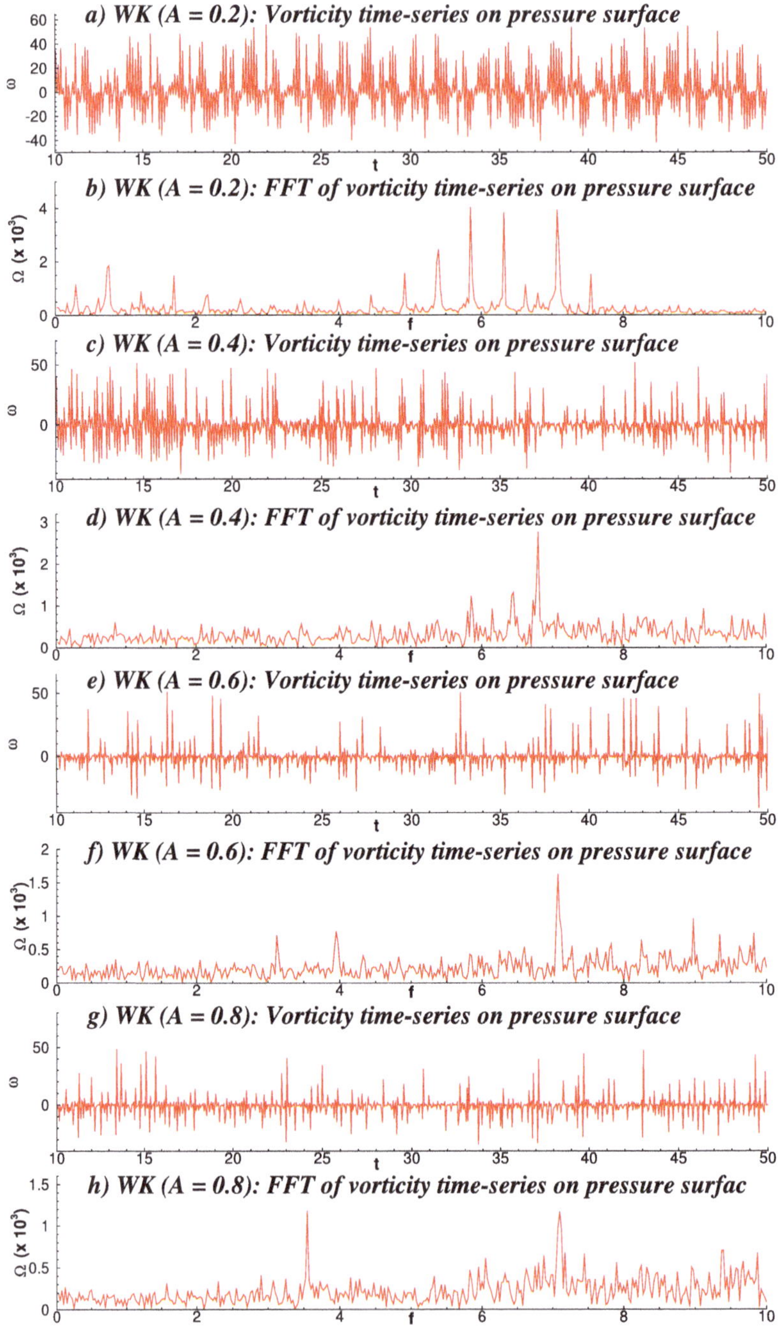

Fig. 5 Time series of vorticity (**a, c, e, g**) and its FFT (**b, d, f, h**) on the suction surface of T106A for cases with wake amplitude: 0.2, 0.4, 0.6 and 0.8

shown in Fig. 5c, the time series becomes more turbulent, which is also evident in the spectrum of Fig. 5d. Here, the amplitude of the vorticity spectrum decreases, but the vorticity distribution spans a broader range of frequencies. Elevated vorticity is observed at lower frequencies, indicating that increased wake amplitude affects vortex shedding on the pressure surface, triggering a more multi-periodic behavior of the flow. This trend continues with higher wake amplitudes, as shown in Fig. 5e and g where increased unsteadiness is evident. This fact is corroborated by the spectra in Fig. 5f and h, where a discernible peak in the frequency domain is visibly absent. The amplitude of the spectra decreases, with vorticity redistributed over a wider span of frequencies, indicating a more chaotic and turbulent flow as the wake amplitude increases.

5 Application of Compressible Enstrophy Transport Equation to Explore Role of Wake Amplitude

Enstrophy, which is defined as the scalar product of the vorticity vector (ω) with itself, serves as a quantification of rotational characteristics within a fluid. It is generated and transported within the internal flow of a T106A cascade through various mechanisms. This concept is analogous to the turbulent kinetic energy equation, which measures translational energy within the flow field. Recently, a compressible enstrophy transport equation (CETE) has been formulated by applying the compressible Navier-Stokes equations (NSE) without any simplifications, as detailed in [16]. This represents an extension of the traditional enstrophy transport equation used to track disturbances in an incompressible flow, as discussed in [41]. The CETE has proven valuable in elucidating the initiation of hydrodynamic instabilities induced by acoustic phenomena and baroclinic mechanisms, as documented in [42–44].

In this context, we delve into the individual components of the CETE, examining their relevance in the separation-induced transition occurring on the suction surface of the T106A cascade, for the first time. This analysis aims to elucidate the distinct mechanisms triggered by unsteady wakes impinging at the inflow of a T106A cascade and to assess whether variations in wake amplitude significantly impact the transition process. To provide a better understanding of the CETE derivation, we will briefly outline how it is obtained from the compressible vorticity transport equation, starting with the momentum conservation equation. Starting with the compressible NSE, which is given as follows:

$$\rho\left(\frac{\partial \mathbf{V}}{\partial t} + (\mathbf{V}\cdot\nabla)\mathbf{V}\right) = -\nabla p + \nabla(\lambda\nabla\cdot\mathbf{V}) + 2\nabla\cdot(\mu S) + \rho\mathbf{f}_\mathrm{b} \qquad (13)$$

where the rate of strain tensor is

$$S = \frac{1}{2}\left[\nabla \otimes \mathbf{V} + (\nabla \otimes \mathbf{V})^T\right] \tag{14}$$

The influence of the applied body force is denoted as $\mathbf{f}_b$. To derive the compressible vorticity transport equation, we calculate the curl of Eq. (13) while representing the vorticity vector as $\omega = \nabla \times \mathbf{V}$. This yields the following expression:

$$\begin{aligned}\frac{\partial}{\partial t}(\nabla \times \mathbf{V}) + \nabla \times [(\mathbf{V} \cdot \nabla)\mathbf{V}] = &- \nabla \times \left(\frac{\nabla p}{\rho}\right) + \nabla \times \left[\frac{1}{\rho}\nabla(\lambda \nabla \cdot \mathbf{V})\right] \\ &+ \nabla \times \left[\frac{2}{\rho}\nabla \cdot (\mu S)\right] + \nabla \times \mathbf{f_b}\end{aligned} \tag{15}$$

Using the vector identities (ϕ represents a scalar):

$$\frac{1}{2}\nabla(\mathbf{A} \cdot \mathbf{A}) = (\mathbf{A} \cdot \nabla)\mathbf{A} + \mathbf{A} \times (\nabla \times \mathbf{A})$$

$$\nabla \times (\phi \mathbf{F}) = \phi \nabla \times \mathbf{F} + \nabla \phi \times \mathbf{F}$$

the compressible vorticity equation, given by Eq. (15), takes the following form,

$$\begin{aligned}\frac{\partial \omega}{\partial t} = &\nabla \times (\mathbf{V} \times \omega) + \left[\left(\frac{1}{\rho^2}\right)(\nabla \rho \times \nabla p)\right] - \left[\left(\frac{1}{\rho^2}\right)\nabla \rho \times \nabla(\lambda \nabla \cdot \mathbf{V})\right] \\ &+ \left(\frac{2}{\rho}\right)\nabla \times [\nabla \cdot (\mu S)] - \left(\frac{2}{\rho^2}\right)\nabla \rho \times \nabla \cdot (\mu S) + \nabla \times \mathbf{f_b}\end{aligned} \tag{16}$$

From Eq. (16), one can obtain the CETE by defining the enstrophy as, $\Omega_1 = \omega \cdot \omega$, and thereafter taking the dot product of Eq. (16) with ω to obtain

$$\omega \cdot \frac{\partial \omega}{\partial t} = \omega \cdot \nabla \times (\mathbf{V} \times \omega) + \omega \cdot \left[\left(\frac{1}{\rho^2} \right) (\nabla \rho \times \nabla p) \right] - \omega \cdot \left[\left(\frac{1}{\rho^2} \right) \nabla \rho \times \nabla (\lambda \nabla \cdot \mathbf{V}) \right]$$
$$+ \left(\frac{2}{\rho} \right) \omega \cdot [\nabla \times [\nabla \cdot (\mu S)]] - \left(\frac{2}{\rho^2} \right) \omega \cdot (\nabla \rho \times \nabla \cdot (\mu S)) + \omega \cdot (\nabla \times \mathbf{f_b}) \tag{17}$$

The local acceleration term simplifies to, $\omega \cdot \frac{\partial \omega}{\partial t} = \frac{1}{2} \frac{\partial \Omega_1}{\partial t}$, and the above is rewritten as,

$$\frac{D\Omega_1}{Dt} = 2\omega \cdot [\nabla \times (\mathbf{V} \times \omega)] + (\mathbf{V} \cdot \nabla) \Omega_1 + \left(\frac{2}{\rho^2} \right) \omega \cdot [(\nabla \rho \times \nabla p)]$$
$$- \left(\frac{2}{\rho^2} \right) \omega \cdot [\nabla \rho \times \nabla (\lambda \nabla \cdot \mathbf{V})] + \left(\frac{4}{\rho} \right) \omega \cdot [\nabla \times [\nabla \cdot (\mu S)]] \tag{18}$$
$$- \left(\frac{4}{\rho^2} \right) \omega \cdot (\nabla \rho \times \nabla \cdot (\mu S)) + 2\omega \cdot (\nabla \times \mathbf{f_b})$$

In the case of a conservative body force, the final term on the right-hand side of the equation will not have an impact on the CETE. We can further simplify this equation using the vector identity: $\nabla \times (A \times B) = (B \cdot \nabla)A + A(\nabla \cdot B) - (A \cdot \nabla)B - B(\nabla \cdot A)$. As an illustration, let's consider the first term on the right-hand side of Eq. (18), which is given by

$$\nabla \times (\mathbf{V} \times \omega) = (\omega \cdot \nabla)\mathbf{V} + \cancel{\mathbf{V}(\nabla \cdot \omega)} - (\mathbf{V} \cdot \nabla)\omega - \omega(\nabla \cdot \mathbf{V})$$

The second term on the right-hand side is dropped as the vorticity vector is solenoidal, i.e., $\nabla \cdot \omega = 0$. Thus,

$$\omega \cdot [\nabla \times (\mathbf{V} \times \omega)] = \omega \cdot [(\omega \cdot \nabla)\mathbf{V}] - \omega \cdot [(\mathbf{V} \cdot \nabla)\omega] - (\nabla \cdot \mathbf{V})\Omega_1 \quad \text{and,}$$

$$\omega \cdot [(\mathbf{V} \cdot \nabla)\omega] = \frac{1}{2}[(\mathbf{V} \cdot \nabla)\Omega_1]$$

With the above simplifications, the CETE is given as

$$
\begin{aligned}
\frac{D\Omega_1}{Dt} =& 2\omega \cdot [(\omega \cdot \nabla)\mathbf{V}] - \cancel{(\mathbf{V} \cdot \nabla)\Omega_1} - 2(\nabla \cdot \mathbf{V})\Omega_1 + \cancel{(\mathbf{V} \cdot \nabla)\Omega_1} \\
&+ \left(\frac{2}{\rho^2}\right) \omega \cdot [(\nabla \rho \times \nabla p)] - \left(\frac{2}{\rho^2}\right) \omega \cdot [\nabla \rho \times \nabla (\lambda(\nabla \cdot \mathbf{V}))] \\
&+ \left(\frac{4}{\rho}\right) \omega \cdot [\nabla \times [\nabla \cdot (\mu S)]] - \left(\frac{4}{\rho^2}\right) \omega \cdot (\nabla \rho \times (\nabla \cdot (\mu S)))
\end{aligned}
\tag{19}
$$

The various terms of Eq. (19) are as follows:

$2\omega \cdot [(\omega \cdot \nabla)\mathbf{V}]$: Contribution to enstrophy due to vortex stretching (T1).

$(\nabla \cdot \mathbf{V})\Omega_1$: Enstrophy growth/decay due to compressibility (T2).

$\left(\frac{1}{\rho^2}\right) \omega \cdot [(\nabla \rho \times \nabla p)]$: Contribution from baroclinic term due to misalignment of gradients of pressure and density (T3).

$\left(\frac{1}{\rho^2}\right) \omega \cdot [\nabla \rho \times \nabla (\lambda(\nabla \cdot \mathbf{V}))]$: Contribution due to misalignment of vorticity and bulk viscosity gradients (T4) .

$\left(\frac{1}{\rho}\right) \omega \cdot [\nabla \times [\nabla \cdot (\mu S)]]$: Diffusion of enstrophy due to viscous action (T5).

$\left(\frac{1}{\rho^2}\right) \omega \cdot (\nabla \rho \times (\nabla \cdot (\mu S)))$: Contribution due to misalignment of gradients of density and divergence of viscous stresses (T6).

Figure 6 illustrates the evolution of the fractional contributions of various terms in the CETE (T1-T6) along the suction surface of the T106A blade for four different simulated wake amplitudes. Since this analysis involves solving the 2D NSE, the contribution from vortex stretching (T1) is absent. Term T2 represents enstrophy generation due to compressibility effects, with the most substantial fractional contribution occurring in the case with the lowest wake amplitude. This implies

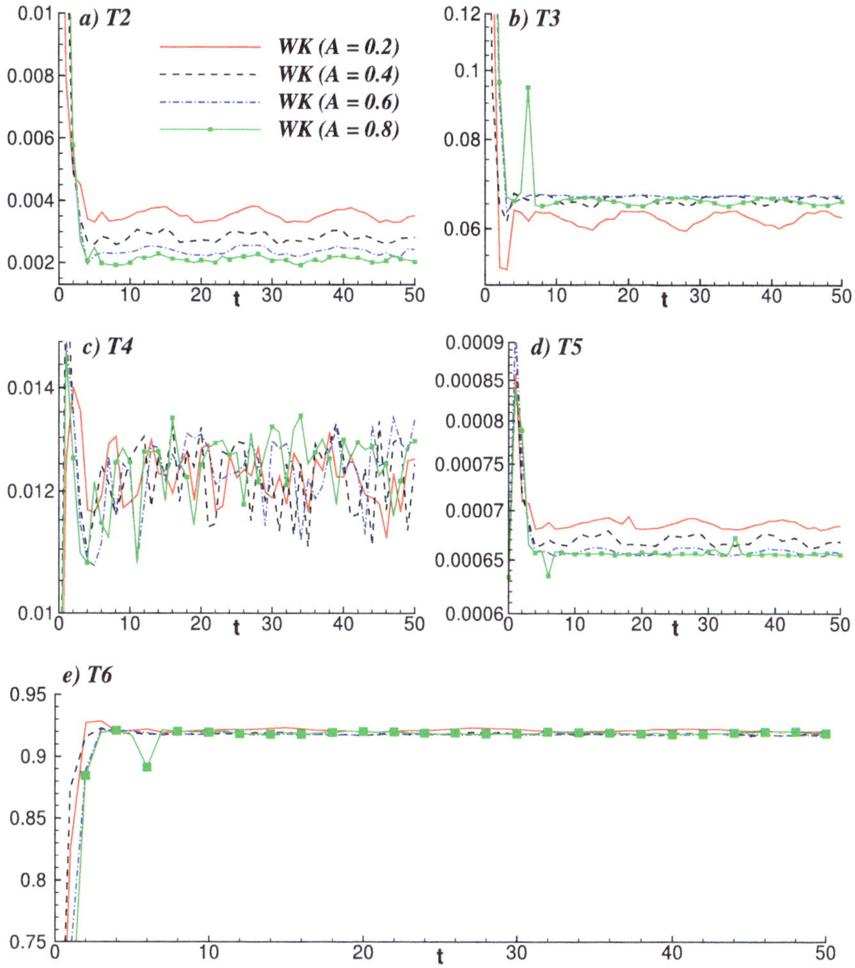

Fig. 6 Evolution of fractional contribution of CETE budget terms (T1-T6) (a-e) on the suction surface of T106A LPT cascade for test cases with wake amplitudes: 0.2, 0.4, 0.6, and 0.8

that increasing the level of free-stream disturbance by raising the wake amplitude has minimal to no impact on the flow's compressibility. Term T3, arising from the misalignment between density and pressure gradients, serving as a source of baroclinic vorticity, exhibits a higher fractional contribution for higher wake amplitudes. However, this effect is primarily observed during the early stages, in the transitional regime of unsteady separation. As the flow becomes turbulent, the influence of wake amplitude on the baroclinic contribution becomes less significant, although there is a general increasing trend with wake amplitude. This can be attributed to the greater number of unsteady separation bubbles on the suction surface as wake amplitude increases, resulting in an overall rise in enstrophy. Term T4 does not exhibit a strong correlation with wake amplitude, showing similar orders of magnitude and trends across all test cases. Term T5, associated with enstrophy diffusion due to viscous effects, has a larger contribution to lower wake amplitudes. As wake amplitude increases, more separation bubbles are observed, each with a lower height than those seen at lower wake amplitudes. Consequently, with increasing wake amplitude, the separation region is suppressed, and the boundary layer thickness increases. This intensifies the influence of viscous action within the shear layer. Term T6, representing the source of baroclinic enstrophy stemming from the misalignment between density gradients and viscous stresses, does not exhibit a strong dependence on wake amplitude. For all cases, regardless of wake amplitude, the largest overall contribution comes from the viscous term (T6), followed by the baroclinic term associated with unsteady separation eddies. This is consistent with prior application of the enstrophy transport equation to hydrodynamic instabilities [42, 43] and for external pressure gradient dominated flows [7, 10]. This suggests that enstrophy creation and distribution are primarily governed by the generation and propagation of eddies within the boundary layer. The unsteadiness is more pronounced in cases with higher wake amplitudes, leading to greater enstrophy generation in those scenarios.

Figure 7 illustrates the time-averaged wall-normal distribution of term T6 from Eq. (19) for the four test cases, obtained by extracting the profile at $S/S_0 = 0.48$. This location is downstream of the peak suction point ($S/S_0 = 0.42$). Notably, as

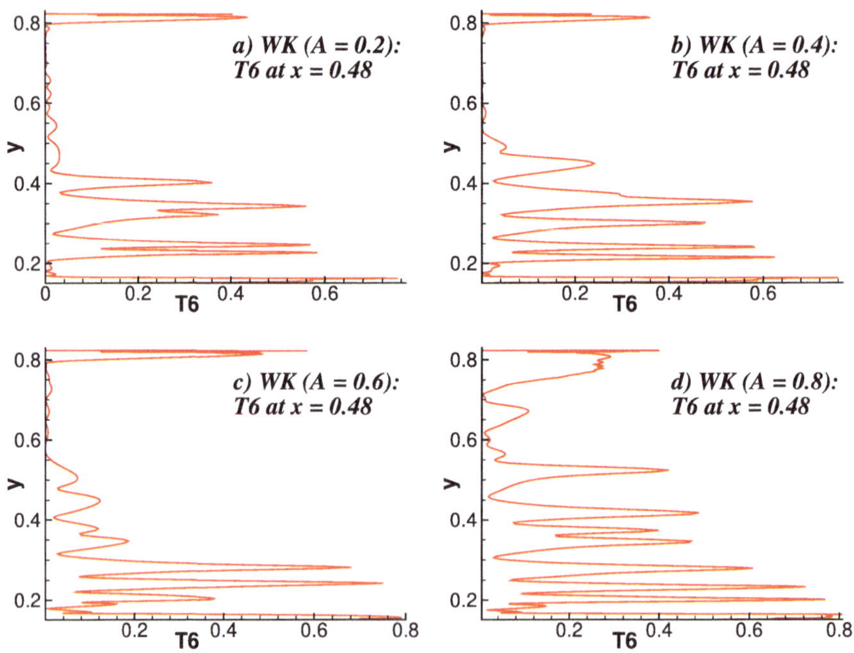

Fig. 7 Time-averaged wall-normal distribution of T6 (extracted at $S/S_0 = 0.48$) compared for cases with unsteady wakes impinging at the inflow with amplitudes: **a** 0.2, **b** 0.4, **c** 0.6 and **d** 0.8

the wake amplitude increases, a larger number of 'peaks' become evident in the distribution. The near-wall peaks correspond to the series of unsteady separations observed along the suction surface, while the upper peak corresponds to vortex shedding events from the pressure surface. Therefore, higher wake amplitudes result in a greater number of unsteady separations on the suction surface, which aligns with the vorticity distribution shown in Fig. 3.

Despite the increased number of separation bubbles observed for higher wake amplitudes, term T6 is distributed over a broader range of wall-normal scales. Consequently, apart from the near-wall peak (which exhibits an increase in T6 magnitude with higher wake amplitude), the other peaks are spread over a larger wall-normal extent and have lower magnitudes of T6. The increased number of peaks also indicates enhanced turbulent mixing in the presence of higher levels of free-stream disturbance, which is a characteristic of separation-induced transition on the suction surface of an LPT blade [5–7]. An analogy can be drawn between the budget terms of enstrophy depicted here and the turbulent kinetic energy equation utilized in previous studies [5, 6]. While only T6, having the greatest contribution to CETE, is displayed, the other terms (T2-T5) also exhibit a series of 'peaks' along the wall-normal extent at $S/S_0 = 0.48$.

Figure 8 presents the time-averaged wall-normal distribution of term T6 from Eq. (19) for the four test cases, with data extracted at $S/S_0 = 0.55$. This location is

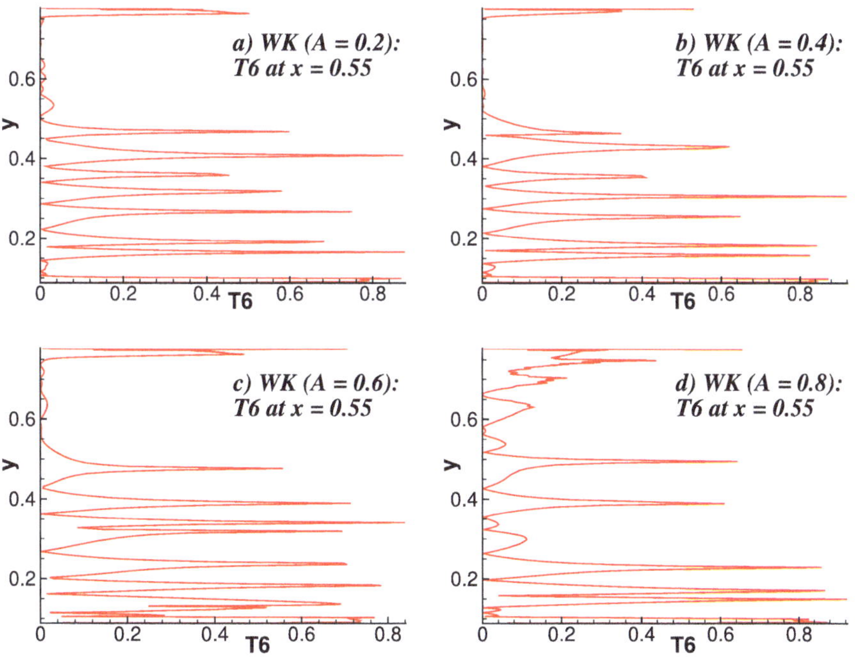

Fig. 8 Time-averaged wall-normal distribution of T6 (extracted at $S/S_0 = 0.55$) compared for cases with unsteady wakes impinging at the inflow with amplitudes: **a** 0.2, **b** 0.4, **c** 0.6 and **d** 0.8

situated beyond the point where flow separation has occurred and is in the transitional regime.

Similar to the previous figure, an increased wake amplitude results in a greater number of 'peaks' in the T6 distribution. At this streamwise location, the contribution from vortex shedding on the pressure side is relatively smaller compared to the near-wall peaks associated with unsteady separations along the suction surface. As the wake amplitude increases, term T6 becomes distributed over a wider range of wall-normal scales, causing the central peaks to spread out and exhibit relatively lower magnitudes compared to cases with lower wake amplitudes. This observation indicates enhanced turbulent mixing at this location compared to its upstream counterpart in Fig. 7. Such behavior is expected as the flow transitions to a turbulent stage, leading to higher mixing rates and enstrophy production [11]. The wake-induced turbulent spots, represented as a sequence of unsteady separations, contribute to a delayed separation of the boundary layer. Additionally, the ensuing calmed regions further suppress the separated flow due to higher shear and fuller velocity profiles. This effect is more pronounced for higher wake amplitudes, as evident in Figs. 7 and 8.

6 Conclusions

The role of wake amplitude on the separation-induced transition in a T106A blade passage (shown in Fig. 1) is investigated by considering four different amplitudes of the Gaussian wake at the inflow: 0.2, 0.4, 0.6, and 0.8. The impact of wake amplitude is first investigated through the instantaneous vorticity distribution in Fig. 3. With an increase in wake amplitude, the thicker boundary layer is noted on the suction surface with a larger number of unsteady separations. The vortex shedding from the pressure surface decreases in intensity with an increase in wake amplitude, as the vorticity is redistributed primarily to the suction surface. The vorticity time series and spectra on the suction surface in Fig. 4, shows enhanced multi-periodicity and unsteadiness with an increase in wake amplitude. Furthermore, for the highest wake amplitude, the frequency of wake passing (2.857 Hz) has the largest amplitude surpassing the Kelvin-Helmhotz vortex shedding frequency, suggesting wake-induced transition being the dominant route for transition, in this case. For the pressure surface in Fig. 5, the vorticity time series and spectra show similar trends as the suction surface vorticity dynamics. With the increase in wake amplitude, the amplitude of the spectrum decreases and the vorticity is redistributed across the frequency plane. Compared to the suction surface, the overall magnitude of vorticity is lower, and even for lower wake amplitudes, there are multiple time periods in the flow field associated with individual vortical eddies of the vortex shedding phenomenon.

One of the discerning features of the present research is the use of a compressible enstrophy transport equation (CETE) to explore the effect of wake amplitude on the generation and distribution of enstrophy in the flow. This equation is derived directly from the compressible Navier-Stokes equation without any assumptions. The study examines various budget terms of the CETE along the suction surface of the T106A blade for different wake amplitudes. The study reveals that wake amplitude has a significant impact on the baroclinic contribution to enstrophy, particularly in the transitional regime. The viscous term is identified as the largest contributor to enstrophy, followed by the baroclinic term associated with unsteady separation bubbles. Higher wake amplitudes lead to increased enstrophy generation and affect the contribution of various terms in the enstrophy transport equation. Enhanced turbulent mixing is also observed by increasing the wake amplitude, which is consistent with prior findings on wake-induced transition [11]. These findings provide valuable insights into the complex flow dynamics and mechanisms involved in separation-induced transition, shedding light on the behavior of boundary layer flow under different wake amplitude conditions.

In the present work, we have shown the effect wake amplitude has on the boundary layer separation on the suction surface of a T106A blade by conducting a parametric investigation using highly accurate numerical methods. It has been found that increasing the wake amplitude and thereby increasing the level of free-stream disturbance leads to a delayed separation and earlier transition, consistent with prior findings of wake-induced transition [10, 11]. The compressible enstrophy budget is evaluated for the first time for wake-induced transition over a T106A blade showing the dominance of the viscous term (T6) and to a smaller extent, the baroclinic term (T3). The smallest contribution has been attributed to the term T5, which is the term involving bulk viscosity. Using Stokes' hypothesis, it is expected that this term will have the lowest fractional contribution. Furthermore, with increasing wake amplitude, greater turbulent mixing is achieved which is visualized in the form of the increased number of peaks in the wall-normal distribution of T6. In the future, we intend to extend our study on the effect of wakes on the separation-induced transition on the suction surface by considering other LPT blade profiles, such as T106B and T106C. Further, we intend to perform a detailed parametric investigation for the effect of wake frequency on the ensuing bypass transition.

Competing Interests This study was funded by the Department of Science and Technology (DST), Government of India, Grant No. SRG/2022/0079. The authors have no conflicts of interest to declare that are relevant to the content of this chapter.

Acknowledgements The author is grateful for the computing facilities (ARYABHATA HPC) provided by IIT Dhanbad for computing the cases presented here. The financial support provided by the Department of Science and Technology (DST), Government of India via Grant No. SRG/2022/0079 is also gratefully acknowledged.

References

1. D.E. Halstead, D.C. Wisler, T.H. Okiishi, G.J. Walker, H.P. Hodson, H.W. Shin, Boundary layer development in axial compressors and turbines: part 1 of 4 - Composite picture (1997)
2. M.R. Banieghbal, E.M. Curtis, J.D. Denton, H.P. Hodson, I. Hunstman, V. Schulte, N.W. Harvey, A.B. Steele, Wake passing in LP turbine blades, in *AGARD Conference Proceedings AGARD CP* (AGARD, 1996), p. 23

3. A. Licari, K. Christensen, Modeling cumulative surface damage and assessing its impact on wall turbulence. AIAA J. **49**, 2305–2320 (2011)
4. J.P. Bons, A review of surface roughness effects in gas turbines (2010)
5. A. Sengupta, N. Vadlamani, P.G. Tucker, Roughness induced transition in low pressure turbines, in *55th AIAA Aerospace Sciences Meeting* (2017), p. 0303
6. M. Alam, N.D. Sandham, Direct numerical simulation of 'short' laminar separation bubbles with turbulent reattachment. J. Fluid Mech. **410**, 1–28 (2000)
7. A. Sengupta, P.G. Tucker, Effects of forced frequency oscillations and free stream turbulence on the separation-induced transition in pressure gradient dominated flows. Phys. Fluids **32** (2020)
8. J. Wissink, W. Rodi, DNS of a laminar separation bubble affected by free-stream disturbances, in *Direct and Large-Eddy Simulation V: Proceedings of the Fifth International ERCOFTAC Workshop on Direct and Large-Eddy Simulation Held at the Munich University of Technology, August 27–29, 2003* (Springer, 2004), pp. 213–220
9. J. Wissink, DNS of separating, low Reynolds number flow in a turbine cascade with incoming wakes. Int. J. Heat Fluid Flow **24**, 626–635 (2003)
10. A. Sengupta, P.G. Tucker, Effects of forced frequency oscillations and unsteady wakes on the separation-induced transition in pressure gradient dominated flows. Phys. Fluids **32** (2020)
11. J.G. Wissink, W. Rodi, H.P. Hodson, The influence of disturbances carried by periodically incoming wakes on the separating flow around a turbine blade. Int. J. Heat Fluid Flow **27**, 721–729 (2006)
12. A. Sengupta, V.K. Suman, T.K. Sengupta, Direct numerical simulation of vortex-induced instability for a zero-pressure-gradient boundary layer. Phys. Rev. E **100**, 588 033118 (2019)
13. A. Sengupta, N. Gupta, B.N. Ubald, Separation-induced transition on a T106A blade under low and elevated free stream turbulence. Phys. Fluids, 36(2)(2024)
14. A. Sengupta, H.N. Ulloa, B. Joshi, Multi-layer Rayleigh-Taylor instability: consequences for naturally occurring stratified mixing layers. Phys. Fluids **35**(10) (2023)
15. P. Sundaram, V.K. Suman, A. Sengupta, T.K. Sengupta, Effects of free stream excitation on the boundary layer over a semi-infinite flat plate. Phys. Fluids **32** (2020)
16. V.K. Suman, P. Sundaram, J.K. Puttam, A. Sengupta, T.K. Sengupta, A novel compressible enstrophy transport equation-based analysis of instability during Magnus-Robins effects for high rotation rates. Phys. Fluids **34** (2022)
17. P. Stadtmüller, L. Fottner, A test case for the numerical investigation of wake passing effects on a highly loaded LP turbine cascade blade, in *Turbo Expo: Power for Land, Sea, and Air*, vol. 78507 (American Society of Mechanical Engineers, 2001), p. V001T03A015
18. A. Sengupta, P. Sundaram, Compressibility effects on the flow past a T106A low pressure turbine cascade. Phys. Fluids **35**(10), 106110 (2023)
19. R. Ranjan, S. Deshpande, R. Narasimha, Direct numerical simulation of compressible flow past a low pressure turbine blade at high incidence, in *Fluids Engineering Division Summer Meeting*, vol. 46216 (American Society of Mechanical Engineers, 2014), p. 602 V01AT02A010
20. L. De Vincentiis, K. Hurovic, D. Lengani, D. Simoni, J. Pralits, D.S. Henningson, A. Hanifi, Effects of upstream wakes on the boundary layer over a low-pressure turbine blade. J. Turbomach. **145**, 051011 (2023)
21. V. Michelassi, L.-W. Chen, R. Pichler, R.D. Sandberg, Compressible direct numerical simulation of low-pressure turbines - part II: effect of inflow disturbances. J. Turbomach. **137**, 071005 (2015)
22. R. Pichler, V. Michelassi, R. Sandberg, J. Ong, Highly resolved LES study of gap size effect on low-pressure turbine stage, in *Turbo Expo: Power for Land, Sea, and Air*, vol. 50787 (American Society of Mechanical Engineers, 2017), p. V02AT40A006
23. F. Hammer, N.D. Sandham, R.D. Sandberg, Large eddy simulations of a low-pressure turbine: roughness modeling and the effects on boundary layer transition and losses, in *Turbo Expo: Power for Land, Sea, and Air*, vol. 51005 (American Society of Mechanical Engineers, 2018), p. V02BT41A014

24. D. Lengani, D. Simoni, R. Pichler, R. Sandberg, V. Michelassi, F. Bertini, On the identification and decomposition of the unsteady losses in a turbine cascade. J. Turbomach. **141**, 031005 (2019)
25. A. Gross, C.R. Marks, R. Sondergaard, Numerical investigation of freestream turbulence effect on flow through low-pressure turbine cascade, in *AIAA Aviation 2022 Forum* (2022), p. 3412
26. M. Fiore, R. Gojon, G. Saez-Mischlich, J. Gressier, LES of the T106 low-pressure turbine: spectral proper orthogonal decomposition of the flow based on a fluctuating energy norm. Comput. Fluids **252**, 105761 (2023)
27. J.D. Denton, Loss mechanisms in turbomachines, vol. 78897 (American Society of Mechanical Engineers, 1993)
28. P. Sagaut, V.K. Suman, P. Sundaram, M.K. Rajpoot, Y.G. Bhumkar, S. Sengupta, A. Sengupta, T.K. Sengupta, Global spectral analysis: Review of numerical methods. Comput. Fluids 105915 (2023)
29. S.K. Lele, Compact finite difference schemes with spectral-like resolution. J. Comput. Phys. **103**(1), 16–42 (1992)
30. P. Sundaram, A. Sengupta, V.K. Suman, T.K. Sengupta, Non-overlapping high-accuracy parallel closure for compact schemes: application in multiphysics and complex geometry. ACM Trans. Parallel Comput. **10**(1), 1–28 (2023)
31. B. Joshi, A. Sengupta, P. Sundaram, Exploring role of aspect ratio for compressible flow in a rectangular lid-driven cavity with a vertical temperature gradient. Phys. Fluids **35**, 066135 (2023)
32. M.M. Opoka, H.P. Hodson, Transition on the T106 LP turbine blade in the presence of moving upstream wakes and downstream potential fields. Turbo Expo: Power for Land, Sea, and Air **47934**, 1091–1104 (2007)
33. C. Hirsch, Numerical computation of internal and external flows. Comput. Methods Inviscid Viscous Flows **2** (1990)
34. W. Lou, J. Hourmouziadis, Separation bubbles under steady and periodic unsteady main flow conditions. J. Turbomach. **122**, 634–643 (2000)
35. S. Karaca, A.G. Gungor, DNS of unsteady effects on the control of laminar separated boundary layers. Eur. J. Mech. B **56**, 71–81 (2016)
36. A. Sengupta, *Numerical Investigation of Disturbance Environments in Low Pressure Turbines.* PhD thesis (University of Cambridge, UK, 2020)
37. P.G. Tucker, *Unsteady Computational Fluid Dynamics in Aeronautics* (Fluid Mechanics and its Applications (Springer, Netherlands, Springer Series, 2014)
38. V. Schulte, H. Hodson, Unsteady wake-induced boundary layer transition in high lift LP turbines. J. Turbomach. **120**, 28–35 (1998)
39. B.R. McAuliffe, M.I. Yaras, Transition mechanisms in separation bubbles under low-and elevated-freestream turbulence. J. Turbomach. **132**, 011004 (2010)
40. A. Sengupta, P. Sundaram, T.K. Sengupta, Nonmodal nonlinear route of transition to two-dimensional turbulence. Phys. Rev. Res. **2**, 012033 (2020)
41. T.K. Sengupta, P. Sharma, A. Sengupta, V.K. Suman, Tracking disturbances in transitional and turbulent flows: Coherent structures. Phys. Fluids **31**, 124106 (2019)
42. A. Sengupta, P. Sundaram, V.K. Suman, T.K. Sengupta, Three-dimensional direct numerical simulation of Rayleigh-Taylor instability triggered by acoustic excitation. Phys. Fluids **34** (2022)
43. A. Sengupta, B. Joshi, A.K. Verma, Thermally stratified free shear layers: combined Kelvin-Helmholtz Rayleigh-Taylor instability. Phys. Fluids **34** (2022)
44. A. Sengupta, B. Joshi, Effects of stabilizing and destabilizing thermal gradients on reversed shear-stratified flows: combined Kelvin-Helmholtz Rayleigh-Taylor instability. Phys. Fluids **35** (2023)

Improving the Cooling Efficiency of Microchips by Vapor Chamber

Kirill Delendik, Tatiana Baranova, Andrei Chorny, Natalia Kolyago, Olga Voitik, and Yulia Zhukova

Abstract Proposed cooling solution on basis of original vapor chamber with plate-fin heat sink is a promising method of solving the heat dissipation problem in high-power electronic devices. CFD simulation of plate-fin heat sink to predict thermal characteristics of different types of plate-cut finning under two ways of heat input to plate-fin heat sink base (direct input and through a top vapor chamber surface as the heat sink base) was performed. It was found that the heat dissipated by the heat sink with plate-cut finning is 82% of required direct heat input. For the heat sink having plate-cut finning with turn of the cut part of finning, the dissipated heat is about 90% at direct heat input from the heat release source and is more than 99% in case of using the top vapor chamber surface as the heat sink base. The fabrication of thin vapor chamber was done for simulation verification. $75 \times 75 \times 0.81\,\text{mm}^3$ vapor chamber (working fluid—water, case—copper, wick ($70 \times 70 \times 0.05\,\text{mm}^3$)—pillars) was developed and tested. Thermal resistance was 0.13 K/W. The temperature uniformity of vapor chamber was 1.4 °C. Maximum heat transfer capacity was 45.5 W. Vapor chamber had good temperature uniformity. Verification confirmed the legitimacy of using the approach for simulation of convective heat transfer of the heat sink.

Keywords Vapor chamber · Pillar wick · Thermal resistance · CFD · Heat sink · Plate-cut finning · Cooling

1 Introduction

Complex miniaturization of electronic components leads to increase in the power density of chips, limiting the use of traditional methods of heat removal from hard-to-reach point sources. More than 50% of all electronics failures are caused by deficiencies in thermal management [1]. Increasing the operating temperature of

K. Delendik (✉) · T. Baranova · A. Chorny · N. Kolyago · O. Voitik · Y. Zhukova
A.V. Luikov Heat and Mass Transfer Institute of the National Academy of Sciences of Belarus, Brovki Str., 15, Minsk 220072, Belarus
e-mail: delendik@itmo.by

D. Zeidan et al. (eds.), *Computational Fluid Dynamics: Novel Numerical and Computational Approaches*, Infosys Science Foundation Series,
https://doi.org/10.1007/978-981-97-8152-2_4

the device by 10 °C reduces its service life by half [2]. With the transition to 5G technologies, chips consume 2.5 times more energy than 4G, which is associated with high speed (5G reduces signal delay to 1 millisecond—versus 10 milliseconds in 4G networks and 100 milliseconds in 3G networks), a large amount of data transfer and speed (from 1 Gb/s—against the maximum 100 Mb/s in 4G networks). This leads to the need for powerful processors with high heat load in a limited space. The problem of overheating comes to the fore and negates all the advantages of the devices. The commercial deployment of powerful chips creates a demand for more complex and advanced technologies for cooling, complicated by a small form factor (additional antennas for 5G), high heat flux density, and contact with the human.

The use of compact vapor chambers and heat pipes permits solving the problem of thermal stabilization of electronic elements because of their low thermal resistance values and uniform temperature distribution abilities. Heat pipe (HP) and vapor chamber (VC) are heat transfer devices that combine the principles of both thermal conductivity and phase transition for extremely efficient heat transfer [3–9]. VC consists of a working fluid, a wick structure, and a vacuum-tight containment case. It is evacuated and partially filled with working fluid. Liquid moves through wick from a cold condenser zone to a hot evaporator zone where it vaporizes. Then vapor moves to the condenser zone. Liquid condenses, releasing its latent heat in condenser zone. The absence of moving mechanical parts in the designs, the movement of the coolant due to capillary force, the high density of the transferred heat flux with minimal temperature differences, autonomy of operation, minimal maintenance, low specific weight, and quiet operation create advantages over other methods of thermal stabilization.

Traditional HP transports heat in one direction, VC is a thermal spreader that transfers heat from a localized source in multiple directions (Fig. 1). The main function of HP is removing loads from heat-stressed elements, the main function of VC is thermal control and prevention of local overheating.

VC is promising solution for heat spreading and dissipating from electronic devices. It can be attached to the base of a heat sink (HS) when best the cooling efficiency is required. This problem is a critical one since overheating will reduce the performance of electronic devices. Usually, the heat generated by the electronic component is transferred to a HS by conduction, and then to the air by forced convection. Main function of a heat sink is reducing the heat flux density by increasing its surface [10]. However, the thermal performance of conventional aluminum or copper HS is limited [11, 12]. As the operating power increases and VC is used, the

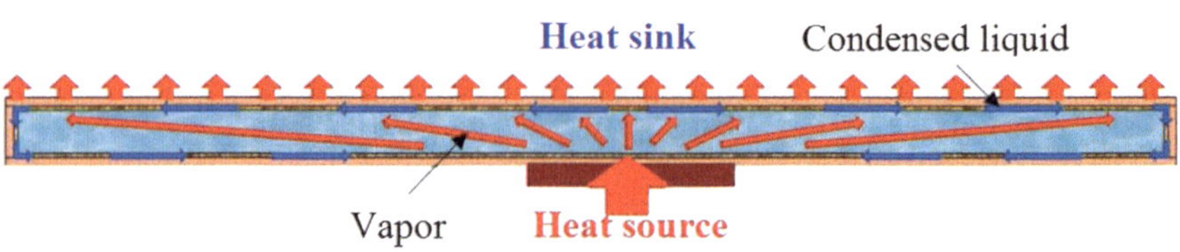

Fig. 1 Vapor chamber operation

generated heat may be transferred more promptly and uniformly, not causing local overheating [13]. VC has a flat contact surface that ensures perfect thermal contact, reduces the contribution of parasitic thermal resistance, and facilitates the layout of the cooling system. The improvement of HS design can be used to achieve electronic cooling performance [10, 11]. One of a such improvement is using a finned HS [11, 14] often having either a plate-fin or pin-fin design [15].

Plate-fin HSs are easier to manufacture whereas pin-fin HSs have better heat dissipation rate but at the expense of additional fabrication cost. A comparison of the thermal performance enhancement of applied finned HSs is presented in [11]. There are the finned HSs designed based on different engineering solutions mainly focused on introducing, e.g., novel fin or splitter configurations (oblique planar [16], flared [17], inclined [18] fins; fins with cutting part [19], with perforation [20, 21], with rectangular hollow [22] or fillet [23] profile, etc.; wavy or arched splitter [24]). For each aforementioned design, many effects have been investigated (fin's material, shape, thickness, orientation, weight, and length; cutting parts by changing the shapes of cutting, numbers, and locations; perforation holes by changing the shapes of holes, locations, diameter, arrangement, and numbers).

The current study is to demonstrate the feasibility of more enhancement in the cooling efficiency of VC with HS with cutting part of plate fins [19]. We made attempts to close this gap and discusses a CFD analysis to introduce efficient cooling systems subject to using VC with special finned HS. Thermal design for plate-fin HS with cutting part was being developed using CFD model (Sect. 2) so that can demonstrate the effect of fin cutting part on HS as a primary phase to increase the thermal performance of conventional plate-fin HS in electronic devices.

Experimental studies of various cooling systems are faced with significant material costs, so the role of numerical modeling becomes extremely relevant. Numerical simulation of gasdynamic processes is based both on the rapid progress, in particular of personal computers (PC) and on the improvement of applied software where the basic principles of computational fluid dynamics, specifically of gasdynamics are realized [25–27, 29–31]. Over the last several decades CFD software has gone the way from primitive scientific programs in the form of separate codes written in different programming languages (FORTRAN, ALGOL, PASCAL, C, etc.) to universal and specialized codes of applied programs. This fact is the result of a wider and successful use of CFD software in different fields:

- aerospace (cooling of gas turbine blades; ventilation and flow in gas turbine cavities; rocket combustion chambers; flame modeling);
- chemical processes (mixers; jet dryers and reactors; porous filler reactors; furnaces, ovens, combustion chambers);
- automobile (external aerodynamics; air movement under the motor hood; cooling system in the car; combustion process in the internal combustion engine);
- mechanical (centrifugal pumps; heat exchangers; recuperators; flowrate meters and pipelines; vapor turbine blades);
- buildings and environment (air heating and conditioning; wind loading; ventilation; fire safety);

- microelectronics (heat sinks and heat transfer enhancement; vapor chambers; electronic installation cooling).

While operating computer processors, controlling chips of different devices, and modern semiconductor circuits release a great amount of heat, which is fraught with such a consequence as a loss of device performance [1, 11, 14, 17]. A working temperature of a microelectronic device is then limited by its heating temperature. Because of small size of the device, it is impossible to control the temperature of the chip itself; its admissible value must be provided by the semiconductor device construction. In practice, the temperature of the semiconductor device shell or the environmental temperature are the parameters to be controlled. The working temperature of a chip is within the range 80–95 °C, it is therefore more often to take 85 °C as the working temperature when designing. It is natural that the effective operation of the shell and the environment requires a lower temperature. Owing to this, the task of effective cooling of semiconductor devices is important and actual.

However, when designing new microelectronics devices sometimes it isn't possible to assess their thermal operating loads. At present, to assess thermal loads on individual elements and, as a whole, on microelectronics devices, the CFD methods are used [25–27, 29–31], allowing one to make studies within the entire thermal and dynamic parameter range, as well as to reveal the necessity of forced cooling of electronics devices and to specify different locations of heat sinks and fans before their physical tests.

Now, universal and specialized software presents complex multicomponent systems that are, as a rule, written in object-oriented programming languages and have a "triad" structure: preprocessor with computational grid builder–solver–postprocessor. Generally, software contains a set of physical and mathematical models for the description of gasdynamic processes, which allows stating a particular task to be solved. However, the computation of complex tasks requires a long enough time, which substantially restricts the software use for CFD. The outcome is to parallelize computational operations and to develop CFD codes for multi-processor systems, which makes it possible to increase many times the speed of solving tasks.

This research serves a platform to improve the thermal efficiency of VC with plate-fin HS with cutting part by introducing new VC and HS designs that offer easier fabrication and implementation. However, to the best knowledge of the authors, the development of a CFD-based model that can demonstrate the effect of cutting part with turn of plate-fin HS to increase the performance of conventional VC with plate-fin HS in electronic devices has not yet been reported. This paper attempts to fill this gap and presents a numerical analysis to compare thermal performance of plate-fin HS with cutting part and those without the cutting part.

Here, the CFD simulation to predict thermal characteristics of different types of plate-cut finning under different heat inputs to a top VC surface as the HS base has been implemented using ANSYS Fluent. ANSYS Fluent is one of the most popular and widespread CFD software and is composed of several codes used for solving individual heat transfer and fluid-dynamics problems. ANSYS Fluent uses the generally accepted approach to choosing special modules. First of all, two flow regimes (steady state and unsteady state) and three basic velocity ranges depending on the medium compressibility (incompressible (low subsonic), weakly compressible (trans-sonic) and strongly compressible (super- and hypersonic) flows) are emphasized. Laminar, transient, and turbulent flow regimes (in terms of the Reynolds number) and separately Newtonian and non-Newtonian flow regimes (in terms of viscosity) are distinguished. Computations take into consideration all modes of heat and mass transfer, including convection and radiation, mixing of chemical components, chemical reactions, and mass transfer.

The purpose of this work was CFD analysis of plate-fin HS to predict thermal characteristics of different types of cutting parts of plate finning under different heat inputs to plate-fin HS with a top VC surface as the HS base and the fabrication of thin VC for verification simulation.

2 Heat Transfer Enhancement of Heat Sink Using CFD Approach

2.1 Governing Equations

For numerical simulation of convective heat transfer in plate-fin HS with a top VC surface as the HS base we used:

Reynolds-averaged steady Navier-Stokes equation

$$\frac{\partial(\rho u_i)}{\partial t} + \frac{\partial}{\partial x_j}\left(\rho \tilde{u}_j u_i - \tau_{ij}\right) = -\frac{\partial p}{\partial x_i} + \rho g_i + S_i \tag{1}$$

continuity equation

$$\frac{\partial(\rho u_j)}{\partial x_j} = 0 \tag{2}$$

energy equations in terms of enthalpy

$$\frac{\partial(\rho h)}{\partial t} + \frac{\partial}{\partial x_j}(\rho \tilde{u}_i h) = \frac{\partial}{\partial x_j}\left[\left(\frac{\mu}{\mathrm{Pr}} + \frac{\mu_t}{\mathrm{Pr}_t}\right)\frac{\partial h}{\partial x_j}\right] + \frac{\partial p}{\partial t} + u_j \frac{\partial p}{\partial x_{j_i}} + \tau_{ij}\frac{\partial u_i}{\partial x_j} \tag{3}$$

In Eq. (3) enthalpy and temperature are related by

$$h = c_p T \tag{4}$$

In Eq. (1) the Reynolds stress tensor τ_{ij} is determined as

$$\tau_{ij} = 2\mu s_{ij} - \frac{2}{3}\mu\frac{\partial u_k}{\partial x_k}\delta_{ij} - \langle\rho u_i' u_j'\rangle \tag{5}$$

In expression (5) the closure term can $\langle\rho u_i' u_j'\rangle$ be presented as

$$\langle\rho u_i' u_j'\rangle = 2\mu_t s_{ij} - \frac{2}{3}\left(\mu_t\frac{\partial u_k}{\partial x_k} + \rho k\right)\delta_{ij} \tag{6}$$

where

$$s_{ij} = \frac{1}{2}\left(\frac{\partial u_i}{\partial x_j} + \frac{\partial u_j}{\partial x_i}\right) \tag{7}$$

The closure problem of the Reynolds-averaged steady Navier-Stokes equation (1) comes down to determining the Reynolds stresses and, as a rule, to writing down additional differential equations for turbulence characteristics. One of the most common models is the model of vortex or turbulence viscosity. The Reynolds equations were closed using the Menter shear stress transport turbulence model [32].

2.2 Problem Statement

The thermophysical characteristics of heat carrier (air) such as density, viscosity, and thermal conductivity were assigned temperature-dependent [33]. As we assumed that the heat sink operated at atmospheric pressure, the pressure-dependent density was ignored. Copper served as the heat sink material.

All computational domain sizes referred to the height of a heat sink fin and were equal to: the distance in front of the heat sink was $6h_r$, behind the heat sink—$25h_r$, and above the heat sink—$2h_r$ (Fig. 2a). To dissipate heat input, heat sinks with different-type plate-cut finning (Fig. 2b) [19, 34, 35] were used. The base area of the heat sink was constant 70 x 70 mm^2; the fin height was h_r = 0.034 m, but the geometry of the finned part was changed. We considered that the plate-cut finning was taken as the basic one. Then we considered the heat sink with plate-cut finning with the turn of the cutting part of the fin. The chosen computational domain was discretized by polyhedral elements with a minimum cell size of $0.183 \cdot 10^{-3}$ m. The total number of the computational domain cells was about 40 mln. Figure 2c illustrates the fragment of the computational domain on the fin surface.

Numerical simulation was performed using the solver Ansys Fluent 2020R1. The numerical simulation used the finite volume method [28]. The values of the incoming flow velocity and temperature T_{in} = 293 K were assigned at the computational domain inlet, and the soft boundary conditions—at its outlet. The considered velocity values corresponded to the turbulent regime of heat carrier flow in the channel. The channel

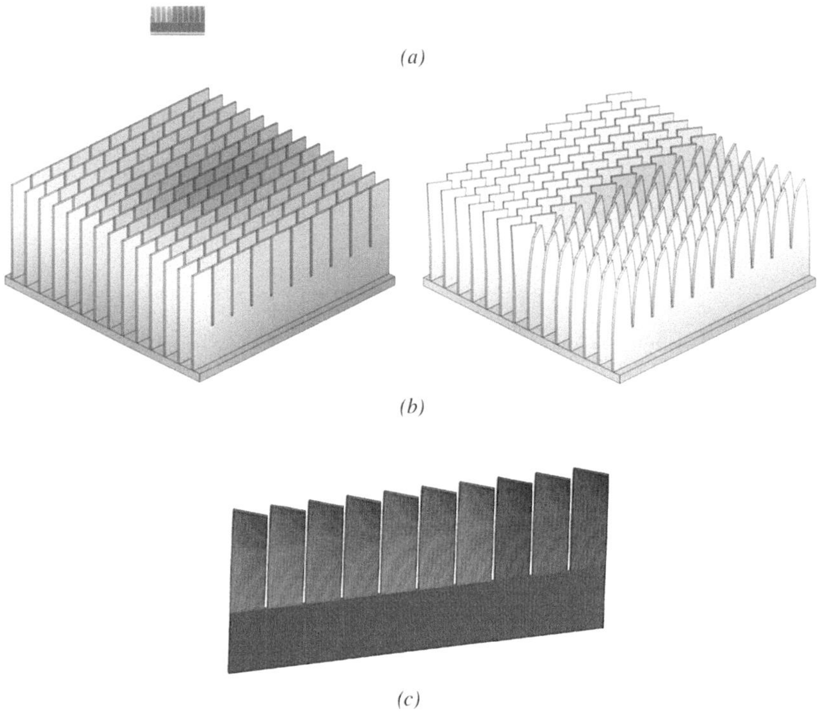

Fig. 2 Computational domain (**a**), heat sink (**b**), and the fragment of the computational domain covering finning (**c**)

walls were assumed to be thermally insulated. The heat flux value was assigned to the heat sink base. Thus, we solved the task in the conjugate statement.

In the heat sink, the fins were soldered to their base with solder, which resulted in a thermal conductivity jump. In the case of numerical simulation, the junction of the heat sink soldered fins and base was modeled using the multiple-layer wall model.

Thus, two-type heat sinks are considered—with plate-cut finning and plate-cut finning with turn of the cut part in order to determine the most effective finning in terms of thermal characteristics. Two ways of heat input to the heat sink base are implemented numerically. In the first case, the 1 x 1 cm^2 area simulating heat input to the heat sink from the heat release source was selected at the base center. In the second case, the heat flux was uniformly distributed over the heat sink base thus simulating the action of the vapor chamber. The amount of heat input to the base was 45 W.

2.3 CFD Results

Figure 3 displays the temperature distribution over the finning surface. In the case of heat removal by the HS with plate-cut finning (Fig. 3a), at direct heat input from the heat release source to the HS base in its central part and on its base, there appears local superheating; with the exception of the frontal part, the HS is heated uniformly. The power dissipated by the HS with plate-cut finning was 36.809 W, which was 82% of input heat. When using the HS with plate-cut finning and turn of the cut part of finning (Fig. 3b) to remove heat at direct heat input from the heat release source to the HS base, the central part is most heated; however, the temperature of the superheated part is lower than that of the HS with plate-cut finning. The fin geometry of the HS with turn of the cut part allows for increasing the vertical component of the flow and removing heat more effectively. The heat dissipated by the HS having plate-cut finning with turn of the cut part was 40.707 W, which was about 90% of the required heat. Thus, the plate-fin HS with turn of the cut part more effectively removes heat in comparison to the HS with plate-cut finning. In the case of uniform heat input to the base of the HS having plate-cut finning with turn of the cut part simulating the action of the vapor chamber, the temperature is redistributed over the entire surface of the HS, and the removed heat is more than 99% of required heat (Fig. 3c).

Requirements for cooling systems of modern electronics are to dissipate significant heat amounts in a confined space effectively. In this space, additional heat release sources can be placed or one cooling system can dissipate heat from two or several sources. Thus, the information on the temperature distribution in a near wake is important from the viewpoint of effective heat removal from the downstream heat release source. Figures 4 and 5 show the temperature and velocity distribution in the wake. In the case of heat removal by the HS with plate-cut finning (Fig. 4a and Fig. 5a) at direct heat input from the heat release source to the HS base, the wake behind the HS is formed with uniform velocity and temperature. In the case of heat removal by the HS having plate-cut finning with turn of the cut part (Figs. 4b, c and 5b, c) both at direct heat input from the heat release source and at uniform heat input to the base of the HS having plate-cut finning with turn of the finned part simulating the action of the vapor chamber, the velocity and temperature distribution in the wake is nonuniform: the temperature wake narrows to the channel center, the central part of the wake has a decreased velocity zone, which allows us to recommend the places of additional heat release sources in more cold wake zones with increased temperature.

3 Development of Vapor Chamber

To verify the numerical simulation of convective heat transfer of the HS, a vapor chamber was developed.

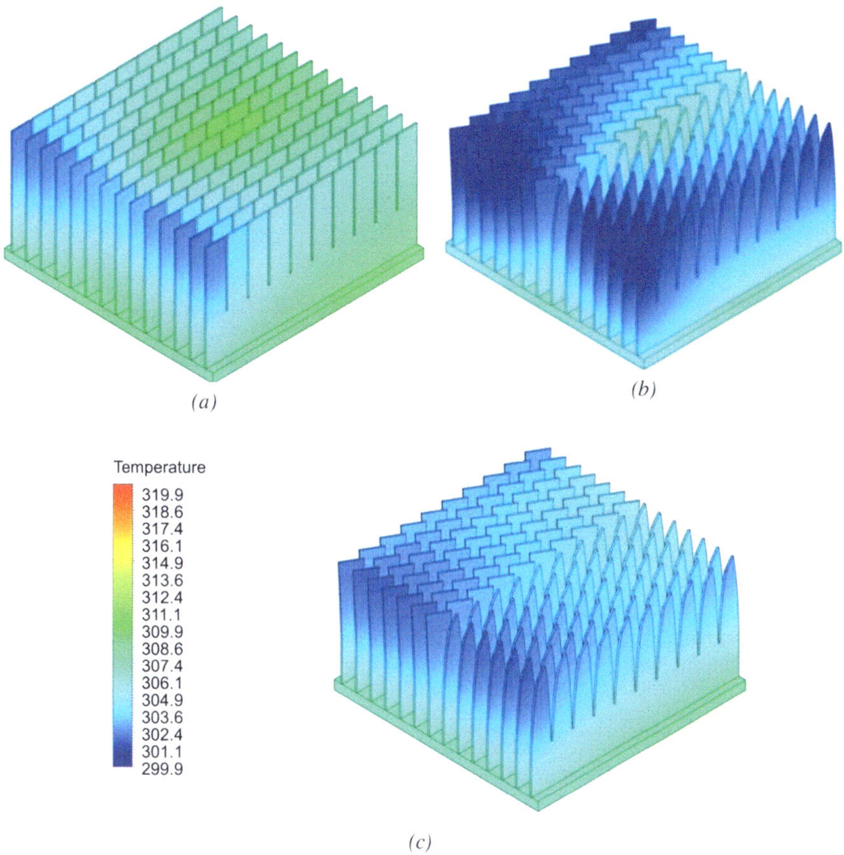

Fig. 3 Temperature distribution over the HS fins: **a**—HS with plate-cut finning; **b**—HS with plate-cut finning with turn of the cut part, direct heat input; **c**—HS with plate-cut finning with turn of the cut part, simulation of the vapor chamber action

3.1 Design of Vapor Chamber

The first stage of development was the choice of materials for the main elements of VC: working fluid, case, and wick [30, 31].

Working Fluid

The working fluid must

- be compatible with the material of wick and case to minimize blockage of VC by non-condensable gases;
- have high latent heat, surface tension, and thermal conductivity;

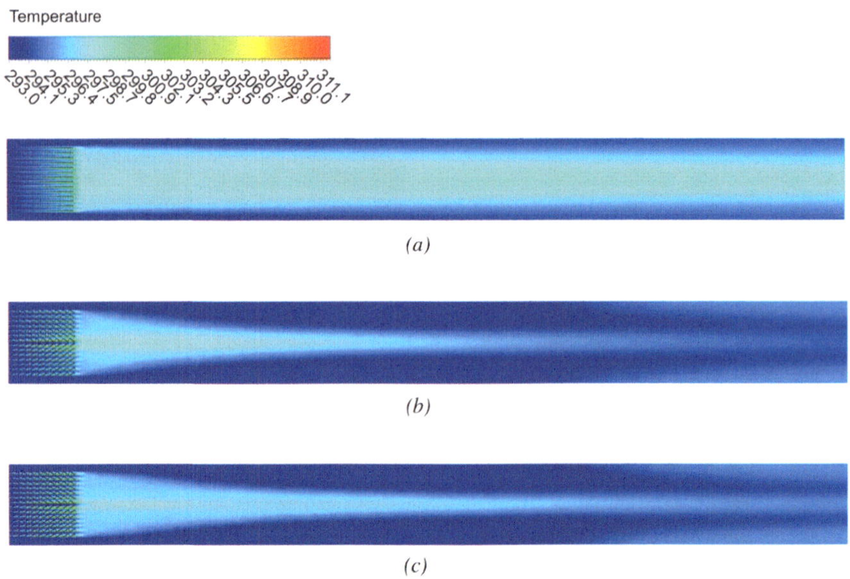

Fig. 4 Temperature distribution in the channel behind the HS in the plane at a height of cuts: **a**—HS with plate-cut finning; **b**—HS with plate-cut finning with turn of the cut part, direct heat input; **c**—HS with plate-cut finning with turn of the cut part, simulation of the vapor chamber action

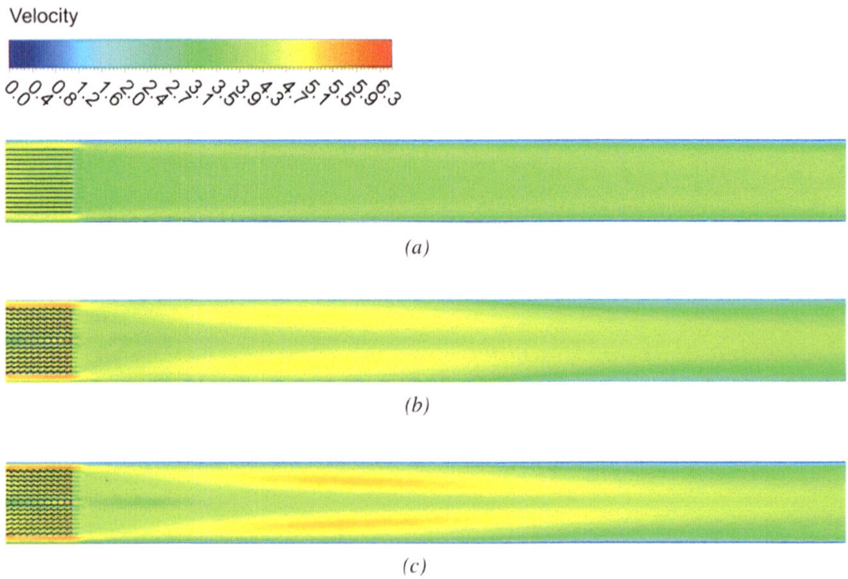

Fig. 5 Velocity distribution in the channel behind the HS at a height of cuts: **a**—HS with plate-cut finning; **b**—HS with plate-cut finning with turn of the cut part, direct heat input;—HS with plate-cut finning with turn of the cut part, simulation of the vapor chamber action

- have a low value of the viscosity of the liquid and vapor phases;
- be safe for humans and environment.

Water selection was based on the merit number that determines the maximum heat transfer capacity of VC, and the Bond number that is the ratio of gravitational forces to capillary forces [51].

Case

The material of the case was copper that passed strict control on material quality, dimensional stability, and tensile and compression strength at temperature and pressure loads during all manufacturing stages. The thickness of the case was 200 ± 3.3 μm that exceeds minimum case thickness determined from the third theory of strength (highest shear stresses) for the maximum saturated vapor pressure. Case parts were etched with a depth of 200 ± 5.2 μm from a copper foil of 400 ± 4.3 μm. The dimensions of the case were 75 × 75 mm.

Wick

There is continuous vapor flow from the evaporator to the condenser and liquid flow from the condenser to the evaporator through the wick at VC operating. The choice of design was determined from limits of maximum heat transfer flux: viscous limit, sonic limit, entrainment limit, boiling limit, capillary limit [5, 31, 36]. The limits have different natures, however, the achievement of any of them leads to an increase in the temperature of the evaporator part and in thermal resistance. The main limits of VC are boiling and capillary.

The capillary structure of the wick is able to provide circulation of liquid up to a certain limit. It occurs when the capillary rate is not sufficient to provide enough liquid to the evaporator zone, the sum of the liquid and vapor pressure drops exceed the maximum capillary pressure that the wick can sustain. Any attempt to increase the heat transfer above the capillary limit leads to drying out in the evaporator zone. The heat flux under the capillary limit was calculated from

$$\Delta p_c \geq \Delta p_l + \Delta p_v + \Delta p_g, \tag{8}$$

where Δp_c—capillary pressure inside the wick structure that necessary for liquid flow back to the evaporator zone; Δp_l—liquid pressure gradient required to return the liquid from the condenser to the evaporation zone; Δp_v—vapor pressure gradient to move the vapor from the evaporation to the condenser zone; Δp_g—hydrostatic pressure drop caused due to the difference in gravitational potential energy.

Capillary pressure Δp_c was calculated using nonlinear partial differential Young-Laplace equation, applying the one-dimensional approximation to the momentum equation liquid pressure gradient; Δp_l and vapor pressure gradient Δp_v were calculated by Chi approach using Poiseuille and modified Blake-Kozeny equations. Δp_g was considered as normal hydrostatic pressure drop.

One way to increase the maximum heat transfer capacity of VC is to use a well-designed wick structure. The purposes of the wick are to generate capillary pressure to transport the working fluid from the condenser to the evaporator and to transfer a sufficient amount of fluid to the evaporator. The maximum capillary drop generated by the wick increases with a decrease in pore size. The wick permeability increases with increasing pore size. The choice of structures of wick leads to trade-offs between high permeability and high capillary capability.

There are various wicking structures: mesh, pillar, grooves, sintered wick, and others. Wicks with a large thin-film area such as pillars are suitable to provide a high heat transfer coefficient [37, 38].

Various pillar geometries (cylindrical, pyramidal, spherical, and conical) were analyzed (Table 1). Structural parameters included shape, characteristic radius r and step of structure S (dimensionless step of the structure $s = S/(2r)$), porosity ε. The porosity was defined as the ratio of void volume to the total volume.

The maximum porosity was possessed by a conical pillar (Fig. 6). Increasing the radius and step of pillars leads to an increase in the effective capillary radius, and as a result, capillary pressure falls. Decreasing the effective radius intensifies heat exchange and reduces the level of superheating corresponding to the beginning of the developed bubble boiling.

It was found that a wick with a cylindrical pillar had high capillary pressure, but low wick permeability (Fig. 7, 8).

It was obtained the values of the dimensionless pillar step to ensure the maximum permeability: for the cylindrical pillar—1.756; for the pyramidal pillar—1.53. For conical and spherical pillars the permeability was increasing function of the pillar step. The most interesting region was $1.35 < s < 1.55$, where the permeability for conical, spherical, and pyramidal pillars differ from each other by no more than 30%. The permeability for cylindrical and pyramidal pillars coincided at $s = 2.881$;

Table 1 Structural and capillary-transport characteristics of different types of pillar [39]

Type of pillar	Porosity ε	Permeability K
Cylindrical	$1 - \frac{\pi}{4s^2}$	$\frac{27.6r^2\varepsilon^{8.43}}{(1-\varepsilon)^{-2.88}}$
Conical	$1 - \frac{\pi}{12s^2}$	$\frac{0.0964r^2\varepsilon^{3.55}}{(1-\varepsilon)^{0.6}}$
Spherical	$1 - \frac{\pi}{6s^2}$	$\frac{4r^2\varepsilon^3}{150\,(1-\varepsilon)^2}$
Pyramidal	$1 - \frac{1}{3s^2}$	$\frac{9.3 \cdot 10^4 r^2\varepsilon^{27.4}}{(1-\varepsilon)^{-4.55}}$

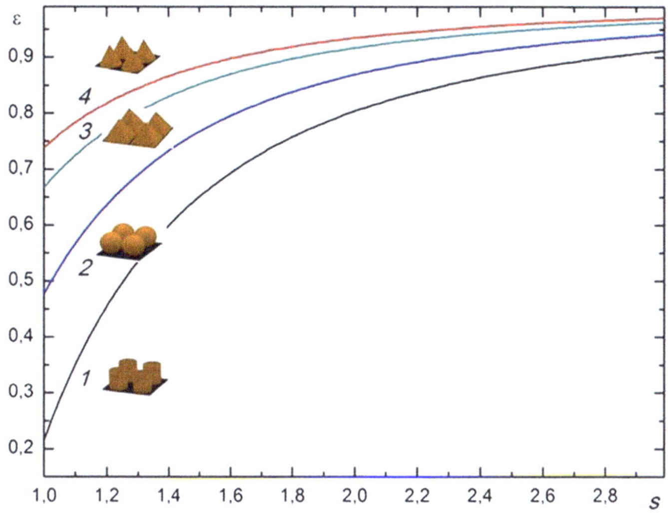

Fig. 6 Dependence of the porosity on the dimensionless step of the structure at r = 60 μm for cylindrical (1), spherical (2), pyramidal (3), conical (4) pillar

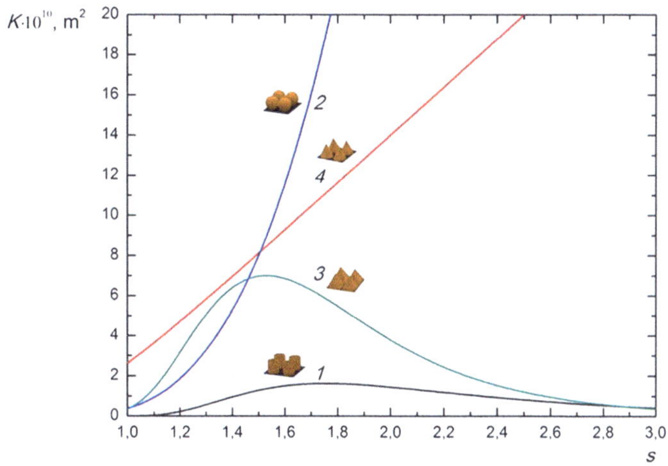

Fig. 7 Dependence of permeability on the dimensionless step of the structure at r = 60 μm for cylindrical (1), spherical (2), pyramidal (3), conical (4) pillar

spherical and pyramidal—at s = 1.016 and s = 1.46; spherical and conical—at s = 1.506. The permeability increased with increasing pillar size. Conical and pyramidal pillars on the surface provided high permeability, good thermal contact, and large thin-film evaporation rates. The spherical pillar structure was promising, but thermal contact with the heat transfer surface for the spherical pillar was extremely small (one point of contact). Although the capillary pressure in the wick with cylindrical

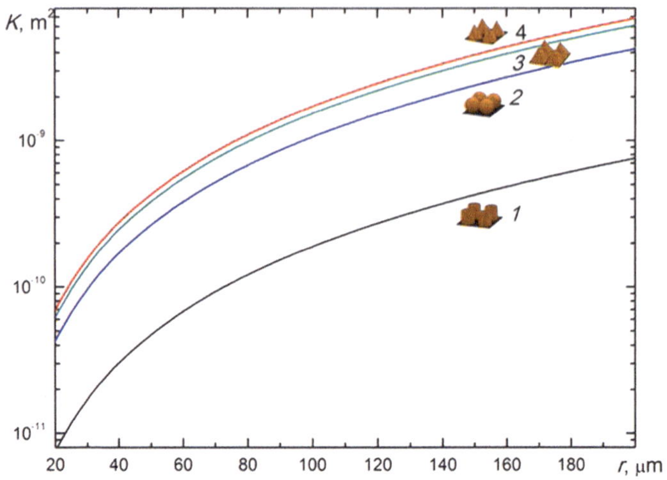

Fig. 8 Dependence of permeability on structure radius at $s = 1.33$ for cylindrical (1), spherical (2), pyramidal (3), conical (4) pillar

pillars was the highest, the heat flux transferred by VC was less due to the greater coefficient of fluid friction. Thus, the wick with conical pillars was characterized by minimum mass (the mass of the wick with conical pillars is three times less than the mass of the wick with cylindrical pillars) with a high efficiency of removing heat loads. It was decided to create the combined type of pillar-conoid. Square packing was chosen.

There are fabrication methods for surface wick structuring such as etching [40], micromechanical cutting [41], micro electrical discharge machining [42], laser processing [43], sintering [44], chemical vapor deposition [45] and 3D printing [46].

Low-cost wet photolithography with utilization of positive photoresist was used. This fact distinguishes developed technology with a difference from similar technologies described in the literature in which expensive dry thick-film photoresists (for example, SU-8) were used. A photomask for photolithography was prepared by the bitmap photoplotter FilmStar XL. The pillar structure was obtained by eco-friendly electrochemical deposition. Figure 9 shows drafts explaining the main stages of the process.

In order to increase hydrophilic properties of wicks methods for their modifying were developed (Fig. 10). The unique morphologies and high surface energy of nanostructures allowed to achieve extreme wettability without introducing large parasitic thermal resistance.

Wick (Fig. 11) was formed for VC by photolithography and electrochemical deposition. Dimensions of wick were 70 × 70 mm (Fig. 12).

Microstructure and pore characteristics were observed by using Mitutoyo Quick Vision Apex (Fig. 13). The top pillars radius was 50.4 ± 2.4 μm (relative error is 5.0% at a confidence level of 0.95), the bottom pillars radius was 150.6 ± 7.8 μm

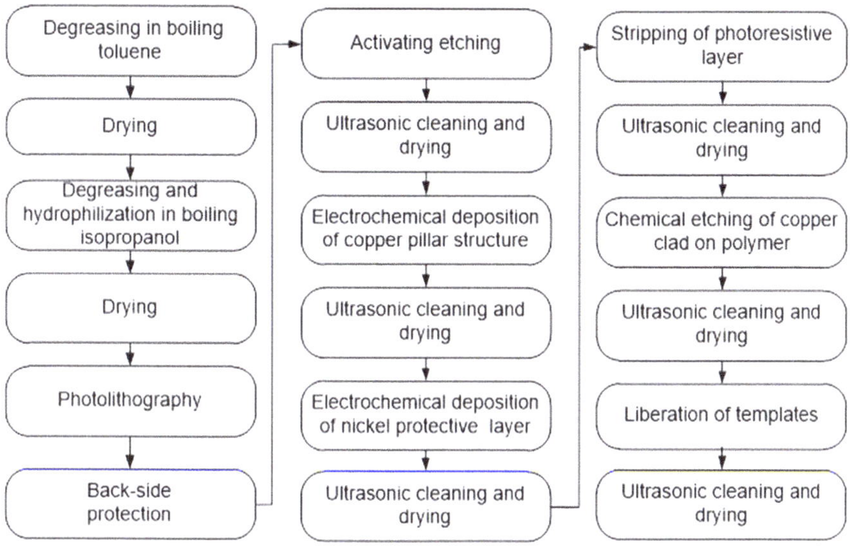

Fig. 9 Block schematic diagram of technology process of wick structure forming

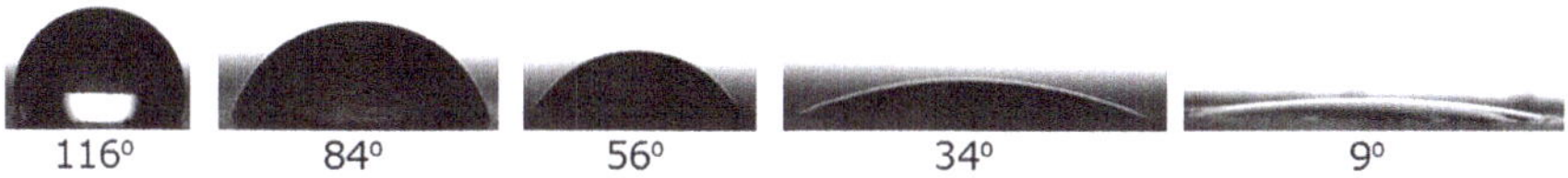

Fig. 10 Change of wetting angle depending on conditions of surface treatment [47]

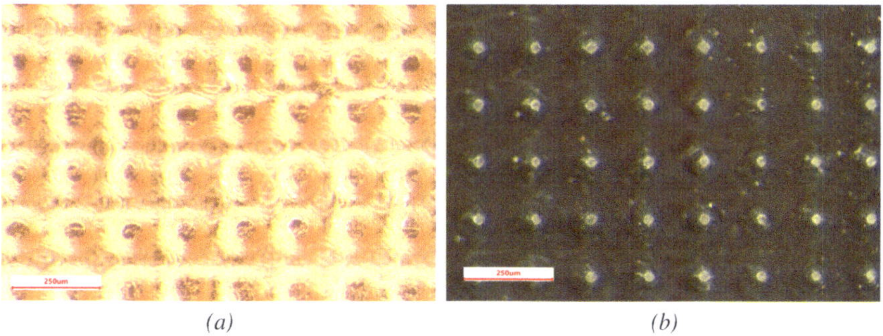

Fig. 11 Initial (**a**) and oxidized (**b**) pillar wick

(relative error is 5.2% at a confidence level of 0.95), the step of pillars was 165.2 ± 13.8 μm (relative error is 8.4% at a confidence level of 0.95), the height of pillars was 48.8 ± 5.6 μm (relative error is 11.5% at a confidence level of 0.95).

Capillary-porous characteristics of wick were investigated by measuring porosity and capillary rise (Fig. 12).

Fig. 12 Photo of oxidized wick sample at the end of capillary rise test

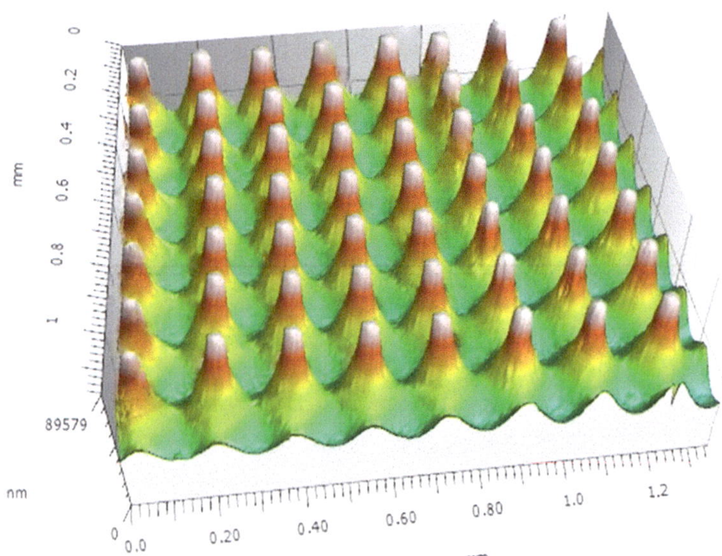

Fig. 13 Microrelief of the pillar wick surface

Porosity measured by the gravimetric method was 65.3 ± 1.6% (relative error is 2.5% at a confidence level of 0.95).

Permeability measured by Darcy-Washburn method was $(9.1 \pm 0.9) \cdot 10^{-11}$ m^2 (relative error 9.9% with a confidence level of 0.95).

Vapor Core

Wick of the condenser and the evaporator parts of the chamber were made in the same way. In the condenser part reinforcing elements were made instead of the pillars. They were used to provide extra mechanical reinforcement for the vapor core and to prevent the collapse of the thin case due to the pressure difference between the interior and exterior of the device, thereby creating the vapor transport core (Fig. 14).

3.2 *Fabrication*

The main manufacturing processes are preparation, photolithography, etching, assembly, welding, charging, and testing.

VC parts were coated with soft solder and then they were sandwiched together by welding. The process of assembling thin chambers is more complex than traditional ones due to burn-through of the thin case, uneven welding of the seam, mechanical damage, and deformation of the wick. Inefficient assembly leads to leakage of the working fluid through defects during operation, which increases thermal resistance and reduces the VC life.

All developed VCs were tested for leakproofness in accordance with State standards 24054-80 and State standards 16493-70 (Table 2).

VC was properly cleaned and evacuated prior to charging to ensure reliable and effective performance. Any contaminates remaining in VC after charging may result in the formation of non-condensable gasses. The presence of non-condensable gases will affect the condensation of water vapor. Non-condensable gases accumulate in the condensation zone, violate the thermodynamic equilibrium at the interface, and can put VC out of operation. Additional treatment of water consists of its boiling in a vacuum for degasation.

There are three main methods of charging techniques: thermodynamic equilibrium method, capillary tubing method, and micro syringe method [49, 50]. The charging set was worked out for evacuation to pressure of approximately 10^{-6} bar and accurate

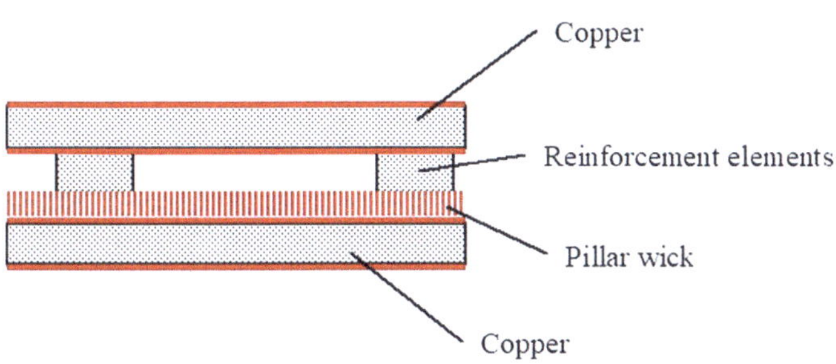

Fig. 14 VC assembly

Table 2 Test procedure for leakproofness

Test procedure	Implementation type	Description	Result
Gas gauge	Compression	VC is filled with argon at an excess pressure of 1.8 atm[a]	No pressure drop was recorded 5 h after cutting off the gas supply
Gas bubble	Compression	VC is filled with argon at an excess pressure of 1.8 atm[a]	Gas bubbles are not recorded
Gas bubble	Heating	VC is filled with argon at an excess pressure of 1.8 atm[a], it's immersed in a bath of distilled water heated to 80 °C	Gas bubbles are not recorded

[a]The pressure of saturated water vapor at 130 °C–270,02 kPa

filling of 0.05-1 ml sized volumes. Basic to this procedure was the micro syringe charging method. The syringe was used to feed the required quantity of fluid into an evacuated VC. Dead volumes in T-joints and other connectors were estimated to predict the actual charged amount.

One of the important VC design characteristics was the dose of charging. For efficient operation of VC, it is necessary that its wick remains saturated with the liquid and the vapor core is filled with vapor. It is important to fill accurately the chamber to avoid underfilling or overflowing with fluid. Excess fluid during VC operation can accumulate in the condenser zone, where a significant temperature gradient occurs, as in the case of the presence of non-condensable gases. This phenomenon does not lead to a complete disruption of operation, but the VC thermal resistance increases significantly. When charging with a lower dose, the heat transfer capacity decreases up to a complete loss of working capacity—dehumidification of the evaporator zone. The optimal dose of charging was 0.91 μl.

After charging VC was sealed (Fig. 15).

The thickness of VC was 0.81 ± 0.02 mm.

3.3 *Thermal Performance*

The thermal performance test was focused on measuring the temperature of VC at the evaporator and condenser zones during the operation of VC for different input powers. During testing, the following requirements were observed: requirements stipulated by the technical documentation for the measuring instruments and equipment used; all necessary measuring instruments and equipment were grounded; experimental equipment were fixed to exclude the possibility of its falling, overturning, and spontaneous displacement during operation.

(a) (b)

Fig. 15 Photos of VC

VC was tested in a horizontal position.

An experimental setup for testing VC was developed. It (Fig. 16) consists of a power supply unit, data acquisition system, test section (Fig. 17), and water cooling system. Wood insulating blocks were set around the assembly to minimize heat loss and also hold the entire assembly in a stable position. The blocks had holes for mounting the thermocouples.

The heating surface was $100\,mm^2$ ($A = 10\,mm \times 10\,mm$). Thermocouple T6 was set close to the heater for the measurement of the evaporator temperature T_e. The heater was capable of providing power up to 100 W. It was powered by a variac with an output range of $U = 0 - 250\,V$ and $I = 0 - 5\,A$.

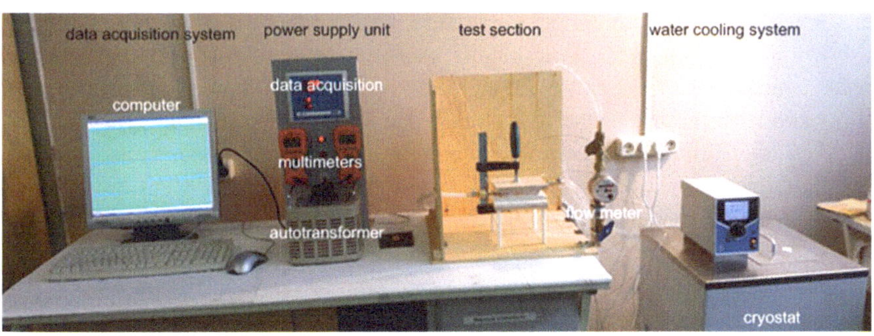

Fig. 16 Experimental setup

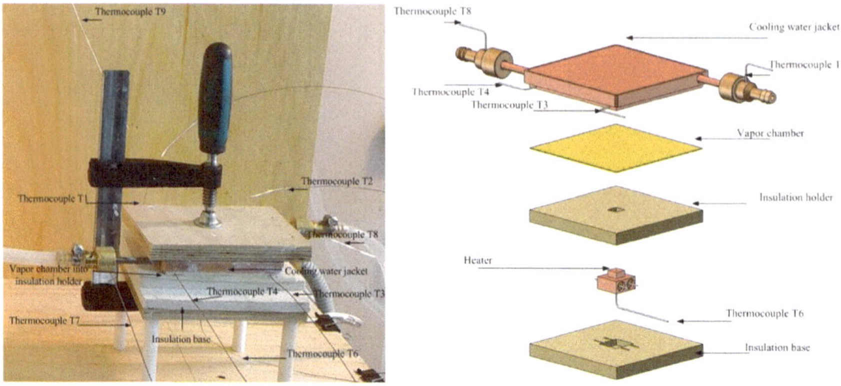

Fig. 17 Test section

A digital power meter was connected to the variac for the measurement of the input power Q_{in}

$$Q_{in} = IU. \tag{9}$$

Applied electrical power was ranged from 1 to 50 W.

A water jacket was set on top of the condenser surface of VC for heat removal. The bottom surface of the cooling water jacket had five tapped holes, which were used to mount the thermocouples. The whole assembly was wrapped in thermal insulation materials, and the heat loss can be thought to be negligible.

To reduce the contact thermal resistance a layer of thermal compound Prolimatech PK-3 with the thermal conductivity of 11 W/(m· K) was smeared on the contact surface between the heater and VC, and the contact one between the water jacket and VC.

Throughout the testing process, the cryostat was kept in the cooling water at a constant flow rate and constant inlet temperature. The distribution of the thermocouples is shown in Fig. 18. Five K-type shielded thermocouples (T1–T5) were attached to the top condenser surfaces of VC for the measurement of the temperature distribution on the condenser side. All the K-type thermocouple measurements were collected by a data acquisition system synchronized to a computer at 0.5 s intervals.

The temperature uniformity of VC was given by the maximum temperature difference on the condenser surface ΔT_c:

$$\Delta T_c = T_{cmax} - T_{cmin}, \tag{10}$$

where ΔT_{cmax} was maximum temperature on condenser part of VC, ΔT_{cmin} was minimum temperature of condenser part of VC. The temperature uniformity of VC was 1.4 ± 0.5 °C.

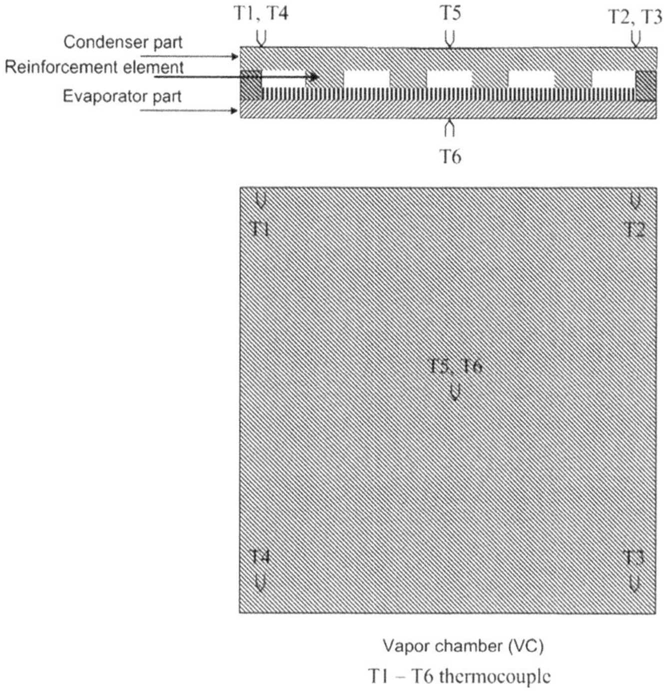

Fig. 18 Distribution of the thermocouples on heat pipe surface

The maximum heat transfer capacity was determined by a sharp jump in temperature between thermocouples or a continuous increase in temperature without establishing steady state. The steady-state condition of VC was considered a state at which changes of temperatures with rate ≤ 1 °C in 5 min. Maximum heat transfer capacity was 45.6 ± 0.9 W.

The thermal resistance of VC characterizes the temperature drop across the device at applied heat load. The thermal resistance of VC R was calculated as

$$R = \frac{T_e - T_{cav}}{Q}, \tag{11}$$

where Q was heat load, W; T_e was the temperature of the evaporator part (T6), °C; T_{cav} was average temperature of the condenser part which can be obtained by

$$T_{cav} = \frac{1}{5}\sum_{i=1}^{5} Ti. \tag{12}$$

The thermal resistance of VC was 0.13 ± 0.01 K/W at 45.6 W. It doesn't account for particular qualities of VC geometry.

Thermal insulation materials guarantee the negligible heat loss that was confirmed by the experimental results: Q_{in} differs slightly from Q_{out} (Q_{out} was approximately 97% of applied power Q_{in}). Thus in the following calculations Q_{in} was used.

Uncertainty of measurements was determined. The source of experimental uncertainty is in all measurements: temperature, voltage, and current. Calibration of thermocouples was carried out with an accuracy of 0.1 °C.

VCs were weighed before and after testing. Damage was not detected after testing.

The thermal resistance is given in Fig. 19 to compare the proposed solution with several VCs reported in other studies. The VCs with different structure sizes and the wick have different maximum heat flux and thermal resistances. The developed VC has competitive characteristics with smaller chamber thicknesses.

Thus VC transferred heat flux 45 W/cm^2 from a localized source with specific power 0.92 W/cm^2.

4 Verification

To verify the numerical simulation of convective heat transfer of the HS, the experimental data of thermal test with vapor chamber and without vapor chamber were used for HS (material—copper; the base area—70 mm × 70 mm × 3 mm; the fin—70 mm × 34 mm × 0.55 mm). The experimental setup was similar to the setup in Fig. 16. Instead of a water cooling system, air cooled heat sink was used. During the

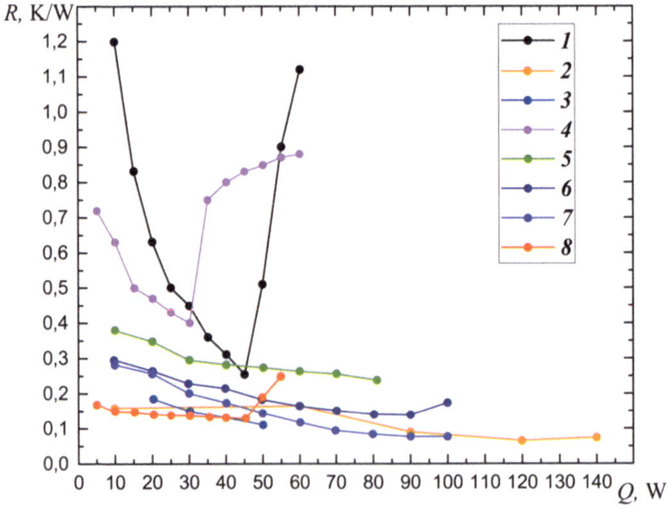

Fig. 19 Comparison of the thermal resistance: *1*—[52], *2*—[53], *3*—[54], *4*—[55], *5*—[56], *6*, *7*—[57], *8*—present data

Table 3 Comparison of theoretical and experimental data

	With vapor chamber		Without vapor chamber	
Value	Experiment	Simulation	Experiment	Simulation
Maximum temperature of heat sink base, °C	63.1	60.9	48.6	47.5
Minimum temperature of heat sink base, °C	58.5	55.9	20.3	20.0

experiment, the environment temperature was maintained at 20±0.3 °C by the air conditioner. The flow rate was set at 2±0.1 m/s. Heat load was 45±0.4 W.

These calculated data have a good correlation with the experimental one (Table 3). The average deviation between simulation and experimental data was 3.2% with a minimum and maximum deviation of 2.3% and 4.4%, respectively. The error lies near the experimental error (3.8%). Verification confirmed the legitimacy of using the approach for the simulation of convective heat transfer of the HS.

5 Conclusions

CFD simulation of plate-fin HS to predict thermal characteristics of different types of plate-cut finning under different heat inputs to plate-fin HS with a top VC surface as the HS base was made. We performed a numerical simulation of convective heat transfer of the HS with different-type plate-cut finning using two ways of heat input (direct input and with a top VC surface as the HS base). We found that the heat dissipated by the HS with plate-cut finning is 82% of the required direct heat input. For the HS having plate-cut finning with the turn of the cut part of finning, the dissipated heat is about 90% at direct heat input from the heat release source and is more than 99% in the case of the top VC surface as the HS base.

The fabrication of a thin vapor chamber was done for simulation verification. 75 mm × 75 mm × 0.81 mm vapor chamber (working fluid—water, case—copper, wick (70 mm × 70 mm × 0.05 mm)—pillars) was developed and tested. Thermal resistance was measured to be 0.13 K/W. The temperature uniformity of VC 1.4 °C. Maximum heat transfer capacity was 45.5 W. VC had good temperature uniformity. Verification confirmed the legitimacy of using the approach for the simulation of convective heat transfer of the HS.

Our studies of convective heat transfer of the HS with different-type plate-cut finning using two ways of heat input allowed obtaining the temperature and velocity distribution in the near wake behind the HS. This permitted us to make recommendations for the installation of additional devices of microelectronics—heat release sources.

Thus proposed cooling solution on the basis of VC is a promising method of solving the heat dissipation problem in high-power electronic devices.

References

1. L.-T. Yeh, *Thermal Management of Microelectronic Equipment* (ASME Press, 2016)
2. Y.A. Cengel, *Heat Transfer: A Practical Approach* (McGraw-Hill, 2003)
3. K.I. Delendik, N.V. Kolyago, O.G. Penyazkov, O.L. Voitik, Development of heat pipes for cooling thermally stressed electronics elements. J. Eng. Phys. Thermophys. **92**, 1529–1536 (2019). https://doi.org/10.1007/s10891-019-02073-8
4. S.U. Khalid, H. Babar, H.M. Ali, M.M. Janjua, M.A. Ali, Heat pipes: progress in thermal performance enhancement for microelectronics. J. Therm. Anal. Calorim. **143**, 2227–2243 (2021). https://doi.org/10.1007/s10973-020-09820-7
5. M. Mochizuki, T. Nguyen, Review of various thin heat spreader vapor chamber designs, performance, lifetime reliability and application. Front. Heat Mass Transf. **13**, 1–6 (2019). https://doi.org/10.5098/hmt.13.12
6. M.J. Gibbons, M. Marengo, T. Persoons, A review of heat pipe technology for foldable electronic devices. Appl. Therm. Eng. **194**, 117087 (2021). https://doi.org/10.1016/j.applthermaleng.2021.117087
7. H. Tang, Y. Tang, Z. Wan, J. Li, W. Yuan, L. Lu, Y. Li, K. Tang, Review of applications and developments of ultra-thin micro heat pipes for electronic cooling. Appl. Energy **223**, 383–400 (2018). https://doi.org/10.1016/j.apenergy.2018.04.072
8. D. Xie, Y. Sun, G. Wang, S. Chen, G. Ding, Significant factors affecting heat transfer performance of vapor chamber and strategies to promote it: A critical review. Int. J. Heat Mass Transf. **175**, 121132 (2021). https://doi.org/10.1016/j.ijheatmasstransfer.2021.121132
9. C.W. Chan, E. Siqueiros, J. Ling-Chin, M. Royapoor, A.P. Roskilly, Heat utilisation technologies: A critical review of heat pipes. Renew. Sustain. Energy Rev. **50**, 615–627 (2015). https://doi.org/10.1016/j.rser.2015.05.028
10. H.E. Ahmed, B.H. Salman, A.S. Kherbeet, M.I. Ahmed, Optimization of thermal design of heat sinks: A review. Int. J. Heat Mass Transf. **118**, 129–153 (2018). https://doi.org/10.1016/j.ijheatmasstransfer.2017.10.099
11. A. Gaikwad, A. Sathe, S. Sanap, A design approach for thermal enhancement in heat sinks using different types of fins: A review. Front. Front. Therm. Eng. **2**, 980985 (2023). https://doi.org/10.3389/fther.2022.980985
12. R.K. Yadav, R. Basak, K.M. Pandey, Review on heat transfer from fins. IOP Conf. Ser. Mater. Sci. Eng. **225**(1), 012145 (2017). https://doi.org/10.1088/1757-899X/225/1/012145
13. Y. Wang, K. Vafai, An experimental investigation of the thermal performance of an asymmetrical flat plate heat pipe. Int. J. Heat Mass Transfer. **43**, 2657–2668 (2000). https://doi.org/10.1016/s0017-9310(99)00300-2
14. Z. Khattak, H.M. Ali, Air cooled heat sink geometries subjected to forced flow: a critical review. Int. J. Heat Mass Tran. **130**, 141–161 (2019). https://doi.org/10.1016/j.ijheatmasstransfer.2018.08.048
15. H.T. Dhaiban, M.A. Hussein, The optimal design of heat sinks: a review. J. Appl. Comput. Mech. **6**(4), 1030–1043 (2020). https://doi.org/10.22055/JACM.2019.14852
16. S.-C. Lin, F.-S. Chuang, C.-A. Chou, Experimental study of the heat sink assembly with oblique straight fins. Exp. Therm. Fluid Sci. **29**(5), 591–600 (2005). https://doi.org/10.1016/j.expthermflusci.2004.08.003

17. Q. Luo, P. Li, L. Cai, X. Chen, H. Yan, H. Zhu, P. Zhai, P. Li, Q. Zhang, Experimental investigation on the heat dissipation performance of flared-fin heat sinks for concentration photovoltaic modules. Appl. Thermal Eng. **157**, 113666 (2019). https://doi.org/10.1016/j.applthermaleng.2019.04.076
18. I. Tari, M. Mehrtash, Natural convection heat transfer from horizontal and slightly inclined plate-fin heat sinks. Appl. Thermal Eng. **61**(2), 728–736 (2013). https://doi.org/10.1016/j.applthermaleng.2013.09.003
19. A.V. Baranyuk, Y.E. Nikolaenko, V.A. Rohachov, A.M. Terekh, P.G. Krukovskiy, Investigation of the flow structure and heat transfer intensity of surfaces with split plate finning. Therm. Sci. Eng. Prog. **11**, 28–39 (2019). https://doi.org/10.1016/j.tsep.2019.03.018
20. G.-J. Huang, S.-C. Wong, C.-P. Lin, Enhancement of natural convection heat transfer from horizontal rectangular fin arrays with perforations in fin base. Int. J. Thermal Sci. **84**, 164–174 (2014). https://doi.org/10.1016/j.ijthermalsci.2014.05.017
21. A.H. AlEssa, M.I. Al-Widyan, Enhancement of natural convection heat transfer from a fin by triangular perforation of bases parallel and toward its tip. Appl. Math. Mech. **29**, 1033–1044 (2008). https://doi.org/10.1007/s10483-008-0807-x
22. U. Akyol, K. Bilen, Heat transfer and thermal performance analysis of a surface with hollow rectangular fins. Appl. Thermal Eng. **26**, 209–216 (2006). https://doi.org/10.1016/j.applthermaleng.2005.05.014
23. K.-C. Wong, S. Indran, Impingement heat transfer of a plate fin heat sink with fillet profile. Int. J. Heat Mass Transfer. **56**, 1–9 (2013). https://doi.org/10.1016/j.ijheatmasstransfer.2013.05.059
24. E. Hosseinirad, M. Khoshvaght-Aliabadi, F. Hormozi, Effects of splitter shape on thermal-hydraulic characteristics of plate-pin-fin heat sink (PPFHS). Int. J. Heat Mass Transfer. **143**, 118586–118601 (2019)
25. H. Towsyfyan, B. Freegah, A.A. Hussain, A.M. El-Deen Faik, Novel design to enhance the thermal performance of plate-fin heat sinks based on CFD and artificial neural networks. Appl. Thermal Eng. **219, A**, 119408 (2023). https://doi.org/10.2139/ssrn.4017030
26. D. Zeidan, P. Bähr, P. Farber, J. Gräbel, P. Ueberholz, Numerical investigation of a mixture two-phase flow model in two-dimensional space. Comput. Fluids **181**, 90–106 (2019). https://doi.org/10.1016/j.compfluid.2018.12.013
27. G. Patankar, J.A. Weibel, S.V. Garimella, On the transient thermal response of thin vapor chamber heat spreaders: Optimized design and fluid selection. Int. J. Heat Mass Transf. **148**, 119106 (2020). https://doi.org/10.1016/j.ijheatmasstransfer.2019.119106
28. S.V. Patankar, Computation of Conduction and Duct Flow Heat Transfer (CRC Press, 1991)
29. Y. Koito, H. Imura, M. Mochizuki, Y. Saito, S. Torii, Numerical analysis and experimental verification on thermal fluid phenomena in a vapor chamber. Appl. Therm. Eng. **26**(14–15), 1669–1676 (2006). https://doi.org/10.1016/j.applthermaleng.2005
30. A. Faghri, Heat pipe: review, opportunities and challenges. Front. Heat Pipes. **5**, 1–48 (2014). https://doi.org/10.5098/fhp.5.1
31. B. Zohuri, *Heat Pipe Design and Technology: Modern Applications for Practical Thermal Management*, 2nd edn. (2016)
32. F.R. Menter, Review of the shear-stress transport turbulence model experience from an industrial perspective. Int. J. Comut. Fluid Dyn. **23**, 305–316 (2009). https://doi.org/10.1080/10618560902773387
33. N.B. Vargaftik, *Handbook on Thermophysical Properties of Gases and Liquids* (Fizmatlit, Moscow, 1963)
34. Y. Nikolaenko, A. Baranyuk, V. Rohachov, A. Terekh, Numerical study of heat transfer and aerodynamic drag of the radiator with lamellar split finning. Arch. Thermodyn. **41**, 67–93 (2020). https://doi.org/10.24425/ather.2020.132950
35. A.V. Baranyuk, V.A. Rogachev, Y.V. Zhukova, A.M. Terekh, A.I. Rudenko, Experimental investigation of heat transfer of plane heat-removing surfaces with plate finning. J. Eng. Phys. Thermophys. **93**, 962–972 (2020). https://doi.org/10.1007/s10891-020-02196-3

36. K. Delendik, N. Kolyago, O. Penyazkov, O. Voitik, Determination of permissible heat fluxes in the heat pipe, in *AIP Conference Proceedings 1978* (American Institute of Physics, 2018), p. 470028
37. S. Bang, S. Ryu, S. Ki, K. Song, J. Kim, J. Kim, Y. Nam, Superhydrophilic catenoidal aluminum micropost evaporator wicks. Int. J. Heat Mass Transf. **158**, 120011 (2020). https://doi.org/10.1016/j.ijheatmasstransfer.2020.120011
38. D. Deng, L. Zeng, W. Sun, A review on flow boiling enhancement and fabrication of enhanced microchannels of microchannel heat sinks. Int. J. Heat Mass Transf. **175**, 121332 (2021). https://doi.org/10.1016/j.ijheatmasstransfer.2021.121332
39. R. Ranjan, A. Patel, S.V. Garimella, J.Y. Murthy, Wicking and thermal characteristics of micropillared structures for use in passive heat spreaders. Int. J. Heat Mass Transf. **55**, 586–596 (2012). https://doi.org/10.1016/j.ijheatmasstransfer.2011.10.053
40. S. Pulak, K. Sanjib, S. Dipak, K.D. Ajoy, B.S. Bidyut, Pool boiling heat transfer and bubble dynamics of modified copper micro-structured surfaces. Int. Commun. Heat and Mass Transf. **134**, 106039 (2022). https://doi.org/10.1016/j.icheatmasstransfer.2022.106039
41. B.Z. Balazs, N. Geier, M. Takacs, J.P. Davim, A review on micro-milling: recent advances and future trends. Int. J. Adv. Manufact. Technol. **112**(3–4), 655–684 (2020). https://doi.org/10.1007/s00170-020-06445-w
42. D. Deng, Y. Xie, Q. Huang, W. Wan, On the flow boiling enhancement in interconnected reentrant microchannels. Int. J. Heat Mass Transf. **108**, 453–467 (2017). https://doi.org/10.1016/j.ijheatmasstransfer.2016.12.030
43. O.J. Allegre, Z. Li, L. Li, Tailored laser vector fields for high-precision micro-manufacturing/CIRP Annals **68**(1), 193–196 (2019). https://doi.org/10.1016/j.cirp.2019.04.125
44. D. Deng, Y. Tang, H. Shao, J. Zeng, W. Zhou, D. Liang, Effects of structural parameters on flow boiling performance of reentrant porous microchannels. J. Micromech. Microengineer. **24**(6), 065025 (2014). https://doi.org/10.1088/0960-1317/24/6/065025
45. T.Y. Liu, P.L. Li, C.W. Liu, C. Gau, Boiling flow characteristics in microchannels with very hydrophobic surface to super-hydrophilic surface. Int. J. Heat Mass Transf. **54**(1–3), 126–134 (2011). https://doi.org/10.1016/j.ijheatmasstransfer
46. T. Dixit, E. Al-Hajri, M.C. Paul, P. Nithiarasu, S. Kumar, High performance, microarchitected, compact heat exchanger enabled by 3D printing. Appl. Therm. Eng. **210**, 118339 (2022). https://doi.org/10.1016/j.applthermaleng.2022.118339
47. O.L. Voitik, K.I. Delendik, N.V. Kolyago, L.Y. Roshchin, Factors influencing the characteristics of wetting of parts of a vapor chamber. J. Eng. Phys. Thermophys. **93**, 1089–1095 (2020). https://doi.org/10.1007/s10891-020-02209-1
48. Y. Nam, S. Sharratt, C. Byon, S.J. Kim, Y.S. Ju, Fabrication and characterization of the capillary performance of superhydrophilic Cu micropost arrays. J. Microelectromechanical Syst. **19**, 581–588 (2010). https://doi.org/10.1109/JMEMS.2010.2043922
49. S. Somasundaram, H. Bin, W. Mengyao, C. Tan, E. Wang, Charging of miniature flat heat pipes. Heat Mass Transf. **54**, 1–6 (2018). https://doi.org/10.1007/s00231-018-2366-7
50. M.T. Ababneh, S. Chauhan, F.M. Gerner, D. Hurd, P. de Bock, T. Deng, Charging station of a planar miniature heat pipe thermal ground plane. J. Heat Transfer. **135**, Is. 2, 021401. https://doi.org/10.1115/1.4007430
51. G. Patankar, J.A. Weibel, S.V. Garimella, Working-fluid selection for minimized thermal resistance in ultra-thin vapor chambers. Int. J. Heat Mass Transf. **106**, 648–654 (2017). https://doi.org/10.1016/j.ijheatmasstransfer.2016.09.038
52. J.S. Chen, J.H. Chou, Cooling performance of flat plate heat pipes with different liquid filling ratios. Int. J. Heat Mass Transf. **77**, 874–882 (2014). https://doi.org/10.1016/j.ijheatmasstransfer.2014.06.029
53. W. Liu, Y. Peng, T. Luo, Y. Luo, K. Huang, The performance of the vapor chamber based on the plant leaf. Int. J. Heat Mass Transf. **98**, 746–757 (2016). https://doi.org/10.1016/j.ijheatmasstransfer.2016.02.091

54. G. Huang, W. Liu, Y. Luo, Y. Li, H. Chen, Fabrication and thermal performance of mesh-type ultra-thin vapor chambers. Appl. Therm. Eng. **162**, 114263 (2019). https://doi.org/10.1016/j.applthermaleng.2019.114263
55. K.V.D. Paiva, M.B.H. Mantelli, L.K. Slongo, Thermal behavior analysis of wire mini heat pipe. J. Heat Transf. **133**(12), 121502 (2011). https://doi.org/10.1115/1.4004526
56. D. Huang, L. Jia, H. Wu, O. Aaker, Experimental investigation on the vapor chambers with sintered copper powder wick. J. Therm. Sci. **30**, 1938–1950 (2020). https://doi.org/10.1007/s11630-020-1366-3
57. X. Meng, S. Tan, Z. Yuan, Y. Zhang, L. Chen, Experimental study on the heat transfer performance of a vapour chamber with porous wick structures printed via metallic additive manufacturing. Int. Communicat. Heat Mass Transf. **140**, 106496 (2023). https://doi.org/10.1016/j.icheatmasstransfer.2022.106496

A Comparative Study of Correlation-Based Transition Models for Aerodynamics in CFD Code NSMB

Abderahmane Marouf, Marianna Braza, and Yannick Hoarau

Abstract In this research article, we present the formulation and implementation of various correlation-based transition models within the framework of an in-house CFD solver, NSMB. The models considered include the zero-equation BCM (Bas-Cakmakcioglu with Modifications) model, the one-equation γ model, and the two-equation γ-Re_θ model, which are coupled with the Spalart-Allmaras and k-ω-SST models. The validation of these models was initially evaluated using the zero pressure gradient flat plate T3A test case, where the turbulence intensity was calibrated for each model. The models were then applied to external aerodynamic low-Reynolds test cases Eppler387 and S809 airfoils, resulting in promising predictions of the aerodynamic lift and drag forces when compared to the fully turbulent approach.

1 Introduction

RANS models are mostly used in the aeronautical research industry because they are more adapted for the design process of aircrafts and their different parts. The fully turbulent hypothesis is addressed in almost every simulation using one-equation Spalart-Allmaras model, [1], and two-equation k-ϵ model, [2], or k-ω SST model, [3], where important mechanisms are neglected such as laminar transition to turbulence. Complex dynamics may occur regarding the flow around wings such as Shock Wave Boundary Layer Interaction "SWBLI", [4], shear-layer instabilities, von Kármán vortex shedding, [5, 6], separated boundary layer due to high angles of attack may lead to a modification of the overall forces as for instance the drag

A. Marouf (✉) · Y. Hoarau
ICUBE Laboratory, University of Strasbourg, Strasbourg, France
e-mail: amarouf@unistra.fr

Y. Hoarau
e-mail: hoarau@unistra.fr

M. Braza
Institut De Mécanique Des Fluides De Toulouse - IMFT, Toulouse, France
e-mail: marianna.braza@imft.fr

D. Zeidan et al. (eds.), *Computational Fluid Dynamics: Novel Numerical and Computational Approaches*, Infosys Science Foundation Series,
https://doi.org/10.1007/978-981-97-8152-2_5

and lift forces, [7]. One of the challenges facing the research community in the field of Computational Fluid Dynamics (CFD) is to predict correctly the flow dynamics and the aerodynamic forces. Despite the progress in Direct Numerical Simulation (DNS) and Large-Eddy Simulation (LES), the transition from laminar to turbulence remains a complex and challenging phenomenon. DNS provides a highly accurate representation of flow dynamics, however, it is computationally intensive and time-consuming. LES, on the other hand, is less demanding compared to DNS in terms of computational resources and has been used in various transition models [8–11] but still expensive. The linear stability method is a useful tool for predicting boundary layer transition, yet it still presents difficulties in three-dimensional scenarios. Different transition models were proposed based on the Reynolds-Averaged Navier-Stokes "RANS" strategy which combines additional transport equations. The study of [12], proposed a single point RANS-based model by combining the simplicity of low-Reynolds number eddy-viscosity models with the capability to capture the production of streamwise fluctuations in the pre-transitional boundary layer and its effect on the bypass transition. This model includes a new three-equation model for the calculation of the mean flow variables such as the velocity, the pressure, and the turbulent fluctuation velocity, as well as turbulent kinetic energy, laminar kinetic energy, and far-field dissipation rate. It has been suggested to incorporate transport equations into RANS equations. The idea can be traced back to early pioneers such as [13–15], who chose to use intermittency as a transport equation. The Local-Correlation based Transition Modeling (LCTM) was developed through the integration of experimental correlations and triggering functions obtained from the transition process, [16]. Two additional transport equations were formulated in this framework corresponding to the intermittency and the transition onset correlation, the concept entitled γ-Re_θ model. This model takes into account turbulence intensity (Tu) in its experimental correlations but does not take into account the Galilean invariance. This implies that its application is restricted to simulations where the transitional walls are in a stationary state relative to the reference frame. This model has gained widespread utilization across a range of industrial domains, from external aerodynamics and turbomachinery flows to a wind turbine, [17], flows and various other industrial applications. Simplified single intermittency transport equation models were recently proposed to avoid the transport equation for Re_θ through the evaluation of the correlation inside the boundary layer, [18], which maintains the LCTM concept of explicit correlations for triggering the transition onset while computing the Re_θ algebraically with local variables. A zero-equation model was proposed by [19], in which an intermittency function was introduced to replace the transport equation. This function serves to mitigate the turbulence production in the Spalart-Allmaras, [1], model through a formulation that is based on local information and empirical correlations, thereby detecting the onset of transition.

The aim of this study is to incorporate various transition models, including zero-equation, one-equation, and two-equation models, based on the LCTM concept into our in-house Navier-Stokes Multi-Block (NSMB) solver (outlined in Sect. 2.1). The models were first calibrated using the flat plate test case of the Ercoftac series (T3A), [20], and then tested for external aerodynamic applications on airfoils Eppler E387,

[21] and S809, [22, 23] at selected Reynolds numbers. The transition including separation bubbles is a critical aspect that is considered in these tests.

2 Numerical Methods

2.1 Flow Solver

The "NSMB", [24] code is used for this study. It solves the Reynolds-Averaged Navier-Stokes equations for compressible fluid flows. NSMB employs a multi-block structured grid strategy. Space discretization schemes, which can be used and include second and fourth-order central schemes with artificial dissipation, as well as various upwind schemes such as Roe, AUSM, [25], and Van Leer, [26], which can be up to fifth order. Time integration can be done using explicit methods like Runge-Kutta schemes or implicit methods like LU-SGS. To improve convergence, techniques like multigrid, local time stepping, preconditioning, and full multigrid can be applied. Additionally, there is a wide range of turbulence models that are widely used in the aeronautical industry, including the one-equation Spalart-Allmaras model, [1], the Organized Eddy Simulation model, [27], and the two-equation model by Menter, [3]. The code contains hybrid RANS-LES approaches, which have been used for many research applications [28–30] and URANS modeling regarding highly detached flows [7, 31, 32].

2.2 Formulation of Transitions Models

2.2.1 BCM Zero-Equation (Algebraic) Transition Model

This transition model is implemented with the Spalart-Allmaras SA model. The BCM (Bas-Cakmakcioglu with Modifications) transition model, [19], involves the modification of the production term in order to facilitate the transition to turbulence. This is can be achieved by multiplying the production term with the intermittency distribution function γ_{BC}:

$$\frac{\partial \nu_T}{\partial t} + \frac{\partial}{\partial x_j}(\nu_T u_j) = \boldsymbol{\gamma_{BC}} C_{b1} S \nu_T - C_{w1} f_w \left(\frac{\nu_T}{d}\right)^2 + \frac{1}{\sigma}\left\{\frac{\partial}{\partial x_j}\left[(\nu_L + \nu_T)\frac{\partial \nu_T}{\partial x_j}\right] + C_{b2}\frac{\partial \nu_T}{\partial x_j}\frac{\partial \nu_T}{\partial x_j}\right\} \tag{1}$$

The intermittency function is defined as

$$\gamma_{BC} = 1 - exp(-\sqrt{Term1} - \sqrt{Term2}) \tag{2}$$

The $Term1$ is defined as

$$Term1 = \frac{max(Re_\theta - Re_{\theta c}, 0.0)}{\chi_1 Re_{\theta c}} \tag{3}$$

The calibration constant χ_1 is equal to 0.002, the momentum thickness Reynolds number and the vorticity Reynolds number are specified as

$$Re_\theta = \frac{Re_{\upsilon,max}}{2.193} \quad and \quad Re_\upsilon = \frac{\rho d_w^2}{\mu}\Omega \tag{4}$$

The critical momentum thickness Reynolds number $Re_{\theta c}$ is a correlation based on transition experiment data, [19], and is defined as

$$Re_{\theta c} = 803.73(Tu_\infty + 0.6067)^{-1.027} \tag{5}$$

Concerning the Term2 is defined as

$$Term2 = max\left(\chi_2 \frac{\nu_t}{\nu}, 0.0\right) \quad \chi_2 = 50 \tag{6}$$

2.2.2 Two-Equation LCTM Model

The present transition model has been developed by [33], after some modifications of the models [16], to improve the prediction of a natural transition. The current transition model is outlined as follows:

The intermittency γ transport equation is defined as

$$\frac{\partial \rho\gamma}{\partial t} + \frac{\partial(\rho U_j \gamma)}{\partial x_j} = P_{\gamma 1} - E_{\gamma 1} + P_{\gamma 2} - E_{\gamma 2} + \frac{\partial}{\partial x_j}\left[\left(\mu + \frac{\mu_t}{\sigma_\gamma}\right)\frac{\partial \gamma}{\partial x_j}\right] \tag{7}$$

The sources' terms of transition are specified as follows:

$$P_{\gamma 1} = F_{length} C_{a1} \rho S [\gamma F_{onset}]^{C_a}; \quad E_{\gamma 1} = C_{e1} P_{\gamma 1} \gamma \tag{8}$$

In the present model, "S" represents the magnitude of the strain rate, and "F_{length}" is an empirical correlation that governs the length of the transition region. The sources of destruction and relaminarization are specified as follows:

$$P_{\gamma 2} = C_{a2} \rho \Omega \gamma F_{turb}; \quad E_{\gamma 2} = C_{e2} P_{\gamma 2} \gamma \tag{9}$$

Ω represents the magnitude of the vorticity. The transition onset is governed by the following function:

$$Re_\upsilon = \frac{\rho y^2 S}{\mu}; \quad R_T = \frac{\rho k}{\mu \omega} \tag{10}$$

$$F_{onset1} = \frac{Re_{\upsilon}}{2.193 Re_{\theta c}}; \quad F_{onset2} = min\left(max\left(F_{onset1}, F_{onset1}^{4}\right), 2.0\right) \tag{11}$$

$$F_{onset3} = max\left(1 - \left(\frac{R_T}{2.5}\right)^3, 0\right); \quad F_{onset} = max(F_{onset2} - F_{onset3}, 0); \quad F_{turb} = e^{-\left(\frac{R_T}{4}\right)^4} \tag{12}$$

The critical Reynolds number $Re_{\theta c}$ is defined when the intermittency starts to increase in the boundary layer. It takes place upstream of the transition Reynolds number $\tilde{R}e_{\theta t}$ and the difference between these two numbers must be derived through an empirical correlation.

$$Re_{\theta c} = \begin{cases} (-396.035.10^{-2}) + (10120.656.10^{-4})\hat{R}e_{\theta t} + (-868.230.10^{-6})\hat{R}e_{\theta t}^{2} + \\ (696.506.10^{-9})\hat{R}e_{\theta t}^{3} + (-174.105.10^{-12})\hat{R}e_{\theta t}^{4}, \quad \hat{R}e_{\theta t} \le 1870; \\ \hat{R}e_{\theta t} - (593.11 + 0.482(\hat{R}e_{\theta t} - 1870.0), \quad 1870 < \hat{R}e_{\theta t}) \end{cases} \tag{13}$$

The correlations for both F_{length} and $Re_{\theta c}$ are dependent on $\tilde{R}e_{\theta t}$.

The intermittency equation constants are defined as

$$C_{e1} = 1.0; \quad C_{a1} = 2.0; \quad C_a = 0.5; \quad C_{e2} = 50; \quad C_{a2} = 0.06; \quad \sigma_\gamma = 1.0; \tag{14}$$

The separation-induced transition is

$$\gamma_{sep} = min\left(2max\left[\left(\frac{Re_{\upsilon}}{3.235 Re_{\theta c}}\right) - 1, 0\right] F_{reattach}, 2\right) F_{\theta t}; \quad F_{reattach} = e^{-\left(\frac{R_T}{20}\right)^4}; \tag{15}$$

$$\gamma_{eff} = max(\gamma, \gamma_{sep})$$

The boundary condition for γ at the wall is zero normal flux while for an inlet γ is equal to 1.0.

The transition momentum thickness Reynolds number $\tilde{R}e_{\theta t}$ transport equation is

$$\frac{\partial(\rho \tilde{R}e_{\theta t})}{\partial t} + \frac{\partial(\rho U_j \tilde{R}e_{\theta t})}{\partial x_j} = P_{\theta t} + \frac{\partial}{\partial x_j}\left[\sigma_{\theta t}(\mu + \mu_t)\frac{\partial \tilde{R}e_{\theta t}}{\partial x_j}\right] \tag{16}$$

The definition of the source term is as follows:

$$P_{\theta t} = C_{\theta t}\frac{\rho}{t}(Re_{\theta t} - \tilde{R}e_{\theta t})(1.0 - F_{\theta t}); \quad t = \frac{500\mu}{\rho U^2} \tag{17}$$

$$F_{\theta t} = min\left(max\left(F_{wake}.e^{-\left(\frac{y}{\delta}\right)^4}, 1.0 - \left(\frac{\gamma - 1/C_{e2}}{1.0 - 1/C_{e2}}\right)\right), 1.0\right) \tag{18}$$

$$\theta_{BL} = \frac{\tilde{R}e_{\theta t}\mu}{\rho U}; \quad \delta_{BL} = \frac{15}{2}\theta_{BL}; \quad \delta = \frac{50\Omega y}{U}.\delta_{BL} \tag{19}$$

$$Re_\omega = \frac{\rho \omega y^2}{\mu}; \quad F_{wake} = e^{-\left(\frac{Re_\omega}{1E+5}\right)^2} \tag{20}$$

The model constants for the $\tilde{Re}_{\theta t}$ equation are

$$C_{\theta t} = 0.03; \quad \sigma_{\theta t} = 2.0 \tag{21}$$

The boundary condition for $\tilde{Re}_{\theta t}$ at a wall is a zero flux condition. The boundary condition for $\tilde{Re}_{\theta t}$ at an inlet must be determined from an empirical correlation that is based on the turbulence intensity at the inlet.

The two-equation model can be coupled with the Spalart-Allmaras (SA), [34], and the k-ω SST model, [33] by modifying the production and destruction terms of their transport equations.

2.2.3 One-Equation LCTM Model

The γ transition model, [18], requires the resolution of only one transport equation for turbulence intermittency, γ, thereby obviating the requirement for the second equation for Re_θ in the γ-Re_θ model. This leads to a decrease in computational demands and eliminates the dependence on velocity "U", rendering the γ transition model Galilean invariant and allowing its application to surfaces in motion relative to the coordinate system. The formulation of the model is straightforward and can be readily modified with a modest number of user-defined parameters. The intermittency equation is defined as:

$$\frac{\partial \rho \gamma}{\partial t} + \frac{\partial (\rho U_j \gamma)}{\partial x_j} = P_\gamma - D_\gamma + \frac{\partial}{\partial x_j}\left[\left(\mu + \frac{\mu_t}{\sigma_\gamma}\right)\frac{\partial \gamma}{\partial x_j}\right] \tag{22}$$

With the constant $\sigma_\gamma = 1.0$.

The transition source term "P_γ" and the destruction/relaminarization term "D_γ" are defined as follows:

$$P_\gamma = F_{length} \rho S \gamma (1 - \gamma) F_{onset}; \quad F_{length} = 100 \tag{23}$$

$$D_\gamma = C_{a2} \rho \Omega \gamma F_{turb}(C_{e2}\gamma - 1); \quad C_{e2} = 50, \quad C_{a2} = 0.06 \tag{24}$$

where S is the strain rate magnitude. The transition onset is governed by the triggering F_{onset} and F_{turb} which operate within the production and destruction terms of the equation. The detailed equations can be found in [18].

$$Re_v = \frac{\rho d_\omega^2 S}{\mu}, \quad Re_{\theta c} = f(Tu_L, \lambda_{\theta L}) \tag{25}$$

The triggering function is based on Re_υ instead of using the momentum thickness Reynolds number Re_θ that is computed from the velocity profile.

The critical Reynolds number $Re_{\theta c}$ is calculated via an algebraic equation defined as following:

$$Re_{\theta c}(Tu_L, \lambda_{\theta L}) = C_{TU1} + C_{TU2}exp[-C_{TU3}T_{uL}F_{PG}(\lambda_{\theta L})] \quad (26)$$

$$C_{TU1} = 100.0, \quad C_{TU2} = 1000.0, \quad C_{TU3} = 1.0 \quad (27)$$

The empirical function $F_{PG}(\lambda_{\theta L})$ is used to sensitize the transition onset and calibrated using the Falkner-Skan series as follows:

$$F_{PG}(\lambda_{\theta L}) = \begin{cases} \lambda_{\theta L} \geq 0; & min(1 + C_{PG1}\lambda_{\theta L}, C_{PG1}^{lim}) \\ \lambda_\theta < 0 & min(1 + C_{PG2}\lambda_{\theta L} + C_{PG3} \quad min[\lambda_{\theta L} + 0.0681, 0], C_{PG2}^{lim}) \end{cases} \quad (28)$$

$$C_{PG1} = 14.68, \quad C_{PG2} = -7.34, \quad C_{PG3} = 0.0, \quad C_{PG1}^{lim} = 1.5, \quad C_{PG2}^{lim} = 3.0 \quad (29)$$

The transition models are coupled with the SA and SST models through a modification of their original transport equations, in a similar way as for the two-equation transition model.

3 Models Validation

3.1 T3A

The ERCOFTAC T3-series[1] include a large spectrum of freestream turbulence intensities and pressure gradients to validate the laminar-turbulence transition via the friction coefficient analysis, assembled by [20]. The designated test case T3A, was experimentally examined by employing a single-wire fluctuation measurement technique for the characterization of u', v', and w' fluctuations. The T3A test case is fixed by an upstream velocity of 5.4 (m/s), a Reynolds number of $5.6.\ 10^5$, upstream turbulence intensity of 3.0 (%), and a zero pressure gradient. A structured multi-block grid comprising approximately 52,000 cells across 12 blocks was used for steady simulations with 2 cells in the span direction. The 1st cell size is 10^{-6} The space discretization was performed using a fourth-order central scheme. Several preconditioning techniques such as the Artificial Compressibility Method, [35] were compared to the absence of preconditioning. No significant impact was observed with respect to the convergence.

[1] https://www.ercoftac.org/downloads/bulletin-docs/ercoftac_bulletin_54.pdf.

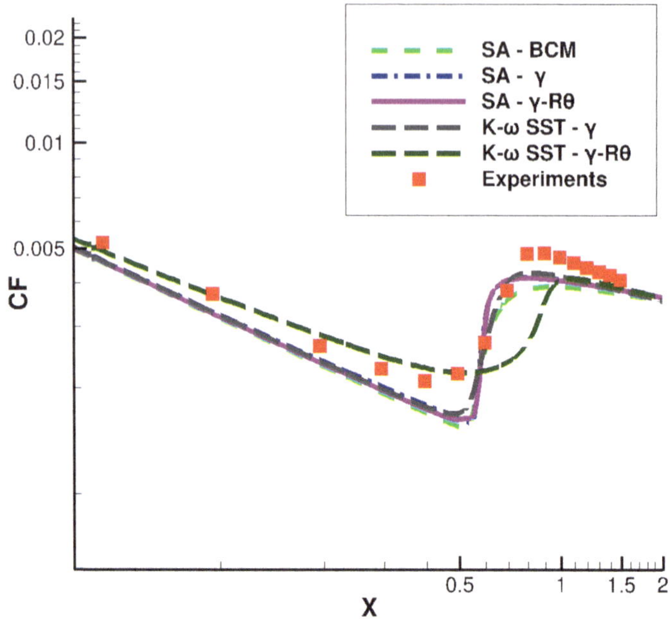

Fig. 1 Comparison of friction coefficients for various transition turbulence models on the flat plate T3A case

The results showed a good agreement with the experimental data for the friction coefficient using various transition models (Fig. 1). The transition onset is approximately similar to the experiments but the length of the transition is underestimated by all the models. The transition in the k-ω-SST - γ-R_θ model captures a smaller transition length with a small delay compared to the other models and to experiments. The sensitivity of transition models to freestream turbulence conditions, such as the turbulence intensity at the inlet, was observed. For the SA model, the sensitivity is related to the initial value of ν_t/ν. This sensitivity can vary depending also on the solver and the numerical method employed for the flow resolution.

3.2 Eppler E387 Airfoil

One example of a two-dimensional validation case is the use of the Eppler E387 airfoil. This case was tested by McGhee and colleagues at the Langley low-turbulence pressure tunnel at a low-Reynolds number of 200,000.

The computational grid used for this test had dimensions of 200,000 cells and was generated as a CH-type grid (Fig. 2) with the first layer set to be 10^{-5} m in order to achieve a y+ less than 1 using a small expansion ratio for a better boundary layer resolution. The freestream turbulence specified for the calculations was 0.1%.

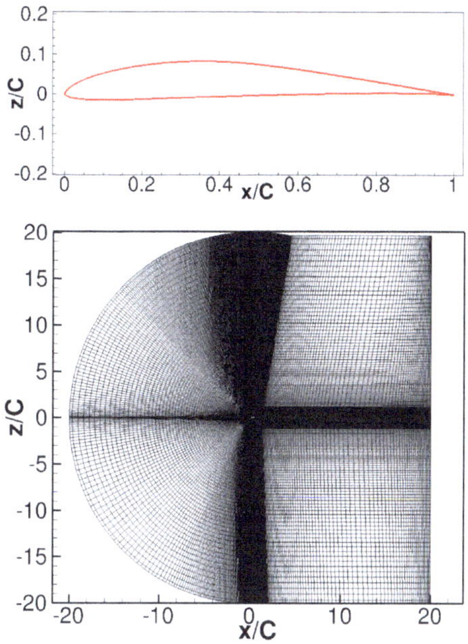

Fig. 2 (Top): Eppler387 airfoil profile. (down): computational domain of the multi-block structured grid

As shown in Fig. 3, the pressure coefficient at different angles of attack is presented. A comparison between the experimental results of [21], different transition models coupled with the Spalart-Allmaras and k-ω-SST, and the fully turbulent standard SA model. The selected transition models are the 0 equation BCM model, the one-equation γ, and the two-equation γ-R_θ. All the models predict well the laminar transition bubble compared to the fully turbulent hypothesis. A slight difference has been noticed for the models in over-predicting the length of the re-circulation. The use of two-equation models such as SA-γ-R_θ revealed a strong correlation with the experimental results. The k-ω-γ-R_θ showed a slightly different transition compared to other models at the angle of 6°. This is due to its sensitivity to the $\mu t_{fs}/\mu_{fs}$ and turbulence intensity. Moreover, the zero-equation model SA-BCM was also able to capture the laminar transition while achieving less CPU time of simulation because of the absence of additional transport equations related to the transition. This can be considerable for three-dimensional applications.

The lift and drag coefficients are both compared with the experimental results in Fig. 4. Both transition models and the fully turbulent model are in good agreement for low-medium angles of attack regarding the lift coefficient (Fig. 4(top)). When this last approaches stall conditions, the fully turbulent model overpredicts the lift because of the under-estimation of the fully separated boundary layer. In the other part, transition models were able to improve the prediction of the lift in this critical region even though simulations employed steady RANS assumption. Figure 4(down) presents the drag polar. The drag coefficient is well predicted when the transition is activated and a good agreement with experimental results is obtained. The drag prediction is a

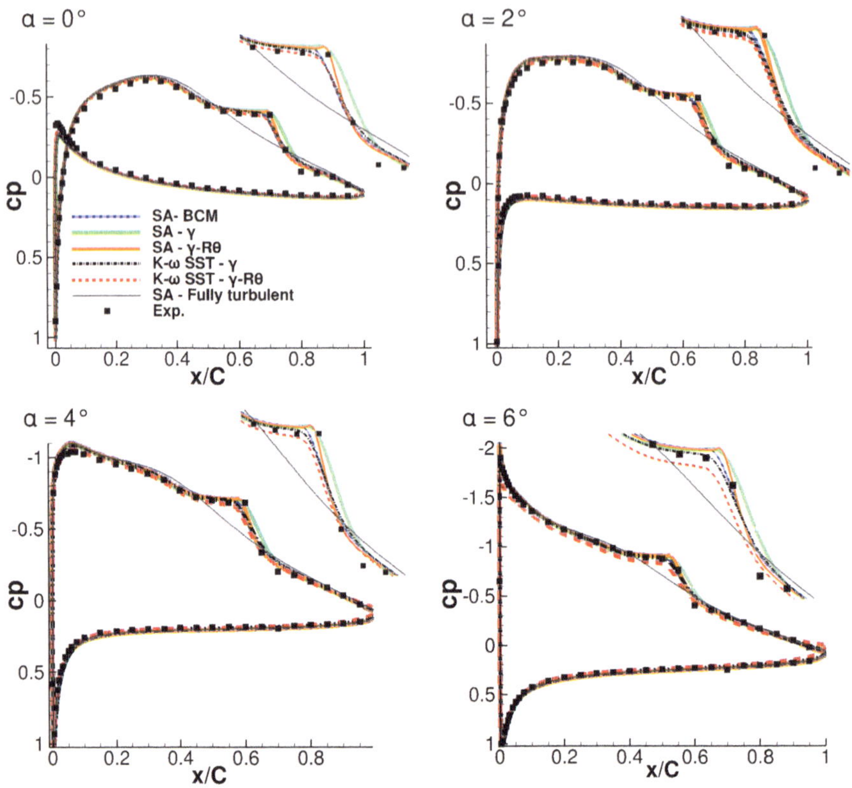

Fig. 3 Comparison of various pressure coefficients (cp) at different angles of attack for different transition models and fully turbulence model

significant challenge such as evidenced in the dictated DPW NASA workshop[2] for improvement of the turbulence modeling for full airplane configuration. In addition, drag reduction constitutes an important challenging task for the improvement of innovative wings at these typical Reynolds numbers where the transition is crucial for the design process. The flow is separated in the laminar region and the laminar separation bubble (LSB) is formed in Fig. 5. The re-circulation develops a certain turbulence that enhances the momentum transport leading to the reattachment of the boundary layer. However, the SA fully turbulent model did not predict any of these mechanisms.

[2] https://aiaa-dpw.larc.nasa.gov/.

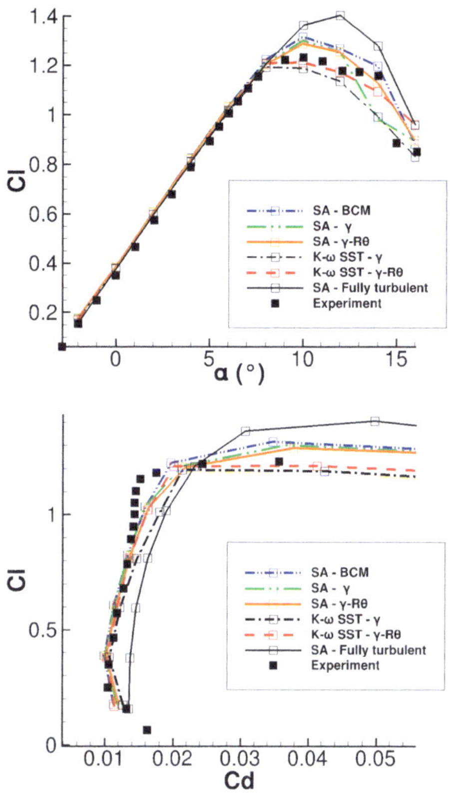

Fig. 4 Comparison of transition models, fully turbulent SA, and experimental results of the Eppler387 airfoil. Lift coefficients (top). Polar drag (down)

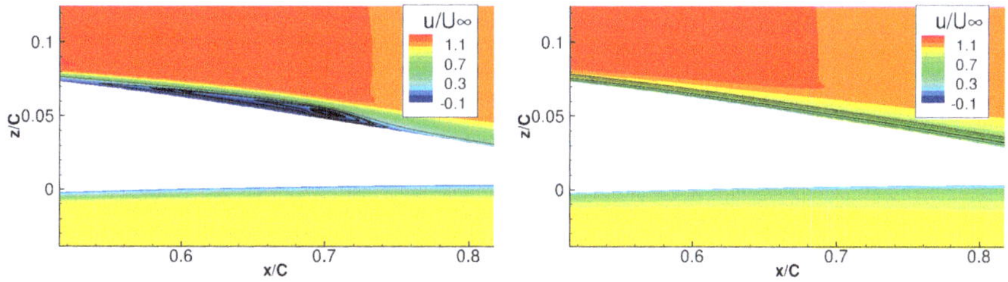

Fig. 5 Non-dimensional u/U_∞ comparison for a transition model SA-γ-Re$_\theta$ (left) and fully turbulent SA model (right)

3.3 *S809 Airfoil*

The S809 wing, specifically designed to be used in horizontal-axis wind turbines, has a thickness of 21% and the profile is designed to have a laminar flow. The experimental results were conducted in the wind tunnel (low-turbulence) at the Delft

University of Technology, [22, 23]. The design process of the airfoil was fixed by two significant objectives. The initial objective was to maintain a non-negative zero-lift pitching moment coefficient, with a minimum value of –0.05. The second objective was the thickness of the airfoil, which was mandated to be 21% of the chord length.

The numerical model presented in Fig. 6 consists in a CH-type 2D structured multi-block grid. Two different grids are used for this study to examine the effects of the mesh sensitivity regarding the transition models. The grid M and F contains 70,000 and 155,000 cells, respectively. The refinement is considered in both horizontal and radial directions. The y^+ is ensured to be < 1. Farfield boundary conditions are considered using 1000 chords. Steady calculations were computed for a Reynolds number of 2×10^6. The freestream turbulence intensity is fixed at the same level as in the low-turbulence wind tunnel corresponding to 0.2% approximately.

The boundary layer remains laminar until 0.5% of the chord. After that, the transition bubble is formed and well captured on both sides of the wing for different transition models as presented in Figs. 7 and 8. Fully turbulent simulations using both SA or k-ω-SST model do not predict the transition and no detachment of the boundary layer is noticed. The transition position is calculated for all the transition models and compared to the experimental results in Fig. 9, and a good agreement has been obtained. However, the existence of a lag has been observed in capturing the transition position at the suction side for certain models such as SA-γ and k-ω-SST-γ. At angles of attack located from 9 to 15°C, the transition on the suction side of the wing is submitted to an adverse pressure gradient, causing this last to move toward the leading edge region.

The lift coefficient is presented in Fig. 10 as a function of the angle of attack using both grids. The results show that transition models are less sensitive to the mesh refinement and a good agreement with experiments is obtained in the range [0–9]

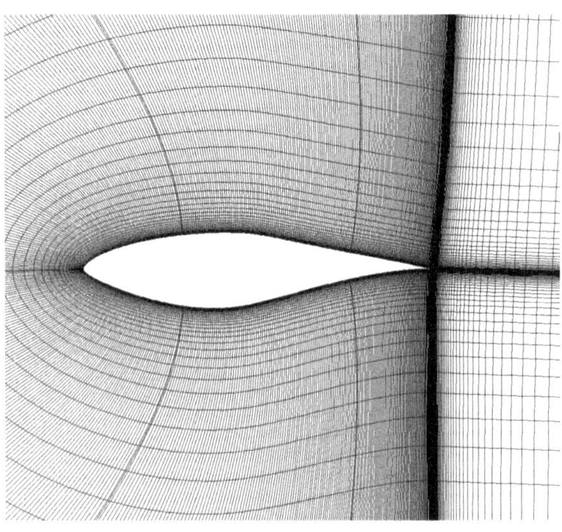

Fig. 6 Computational multi-block structured grid for the S809 airfoil

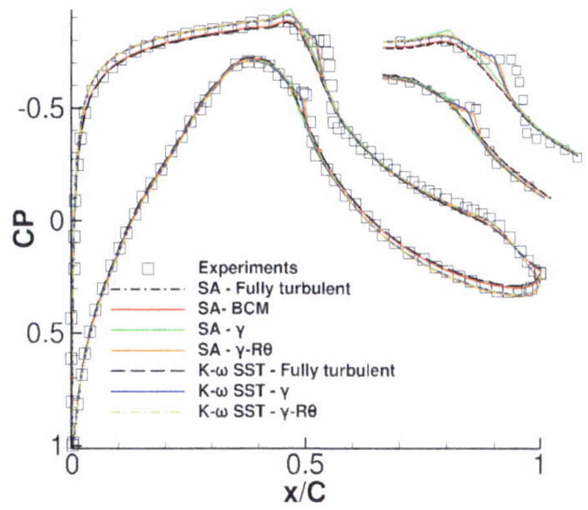

Fig. 7 Comparison of various pressure coefficients (cp) for different transition models and fully turbulence models with experiments at an angle of attack 2.05°C

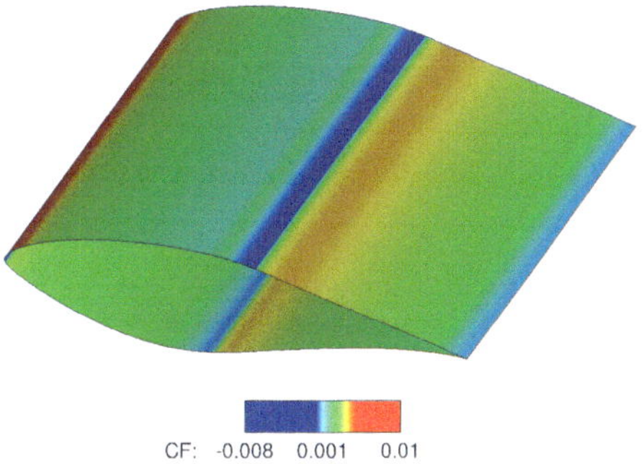

Fig. 8 Friction coefficient (CF) obtained with the transition model k-ω-SST-γ-R_θ at an angle of attack 2.05°C

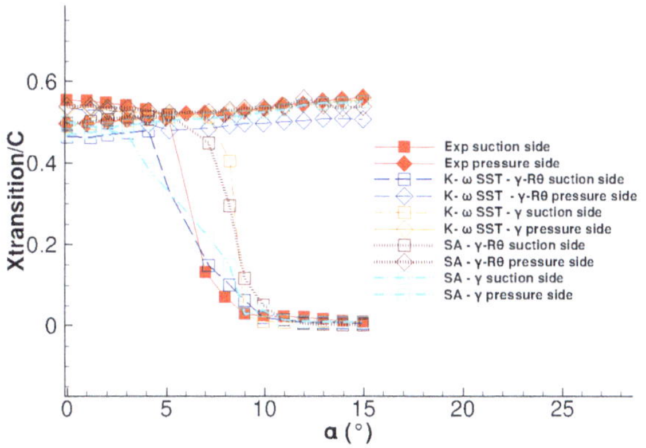

Fig. 9 Transition location at different angles of attack for the suction and pressure sides of the S809 airfoil (grid M)

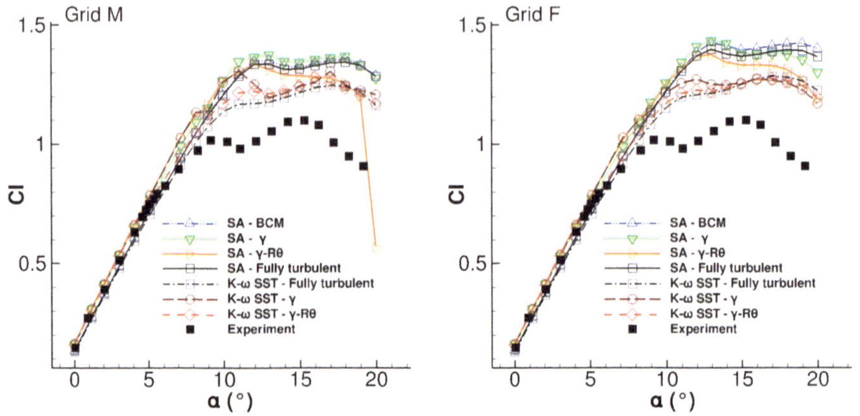

Fig. 10 Lift coefficient comparison of transition models, fully turbulent models, and experimental results of the S809 airfoil for the "Grid M" and "Grid F"

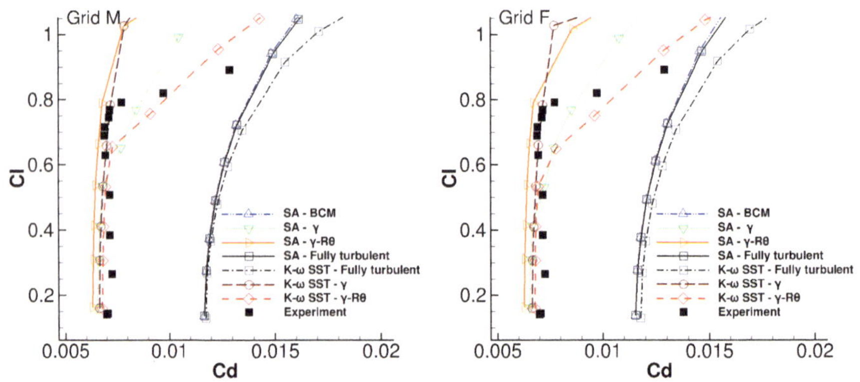

Fig. 11 Drag coefficient comparison of transition models, fully turbulent models, and experimental results of the S809 airfoil for the "Grid M" and "Grid F"

degrees. The drag is well captured by transition models except for the SA-BCM model and fully turbulent models, it is over-predicted by (Fig. 11). At higher angles of attack, the lift is overestimated (as seen in Fig. 10), and the drag is underestimated. The occurrence of a considerable separation near stall conditions in experiments may result in highly unsteady flow dynamics, in addition to the three-dimensional effects of the wind tunnel.

4 Conclusion

Various correlation-based transition models have been implemented into an in-house CFD solver NSMB. The solver uses structured multi-block grids to solve the compressible fluid flows. The BCM zero-equation model was integrated in the Spalart-Allmaras model and does not require any additional transport intermittency equation. In addition, the γ-Re_θ model and its simplified version γ model were implemented and coupled with the SA and k-ω-SST models. Zero pressure gradient flat plate test case T3A was examined for the implemented models. The transition onset and its end are in good agreement with experiments. However, the transition location is highly sensitive to freestream turbulence conditions, such as the imposed turbulence intensity (Tu). On the other hand, transition models have been proven to enhance the accuracy of aerodynamic lift and drag coefficient predictions in steady simulations, compared to using a fully turbulent approach. To summarize the findings of this study, it was observed that two-equation models are more accurate in predicting the onset and length of the transition location compared to other models. However, it should be noted that the effectiveness of these models depends on the specific test case, and this may not always be the case for all scenarios. These models are highly promising and may need further examination in more complex 3D configurations.

Acknowledgements The authors are grateful to the National Supercomputing centres CALMIP, CINES, HPC Strasbourg, TGCC-CEA (A0122A11355) for the Computer allocation, and the PRACE Supercomputing allocation.

References

1. P. Spalart, S. Allmaras, A one-equation turbulence model for aerodynamic flows, in *30th Aerospace Sciences Meeting and Exhibit* (1992)
2. K.-Y. Chien, Predictions of channel and boundary-layer flows with a low-reynolds-number turbulence model. AIAA J. **20**(1), 33–38 (1982)
3. F.R. Menter, Two-equation eddy-viscosity turbulence models for engineering applications. AIAA J. **32**(8), 1598–1605 (1994)
4. J.-B. Tö, N. Simiriotis, A. Marouf, D. Szubert, I. Asproulias, D.M. Zilli, Y. Hoarau, J.C.R. Hunt, M. Braza, Effects of vibrating and deformed trailing edge of a morphing supercritical airfoil in transonic regime by numerical simulation at high reynolds number. J. Fluids Struct. **91**, 102595 (2019)
5. N. Simiriotis, G. Jodin, A. Marouf, P. Elyakime, Y. Hoarau, J.C.R. Hunt, J.F. Rouchon, M. Braza, Morphing of a supercritical wing by means of trailing edge deformation and vibration at high reynolds numbers: experimental and numerical investigation. J. Fluids Struct. **91**, 102676 (2019)
6. A. Marouf, Y. Bmegaptche Tekap, N. Simiriotis, J-B. Tô, J.-F. Rouchon, Y Hoarau, M. Braza, Numerical investigation of frequency-amplitude effects of dynamic morphing for a high-lift configuration at high reynolds number. Int. J. Numer. Methods Heat Fluid Flow **31**(2), 599–617 (2020)
7. A. Marouf, D.H. Truong, Y. Hoarau, A. Gehri, D. Charbonnier, J.B. Vos, M. Braza, Cfd simulations of active flow control devices applied on a cambered flap, in *AIAA SCITECH 2022 Forum* (2022)

8. X. Huai, R.D. Joslin, U. Piomelli, Large-eddy simulation of transition to turbulence in boundary layers. Theoret. Comput. Fluid Dyn. **9**(2), 149–163 (1997)
9. Z. Yang, P.R. Voke, Large-eddy simulation of boundary-layer separation and transition at a change of surface curvature. J. Fluid Mech. **439**, 305–333 (2001)
10. V. Michelassi, J.G. Wissink, J. Frohlich, W. Rodi, Large-eddy simulation of flow around low-pressure turbine blade with incoming wakes. AIAA J. **41**(11), 2143–2156 (2003)
11. P. Catalano, D. de Rosa, Large Eddy Simulations and RANS models for airfoils at low Reynolds number, in *AIAA Aviation 2020 Forum* (2020)
12. D.K. Walters and H.J. Leylek, A new model for boundary layer transition using a single-point RANS approach. J. Turbomachinery **126**(1), 193–202, 03 (2004)
13. J. Steelant, E. Dick, Modeling of laminar-turbulent transition for high freestream turbulence . J. Fluids Engin. **123**(1), 22–30, 10 (2000)
14. Y. Suzen, P. Huang. An intermittency transport equation for modeling flow transition, in *38th Aerospace Sciences Meeting and Exhibit* (2000), p. 287
15. R.B. Langtry, F.R. Menter, Correlation-based transition modeling for unstructured parallelized computational fluid dynamics codes. AIAA J. **47**(12), 2894–2906 (2009)
16. F.R. Menter, R. Langtry, S. Likki, Y. Bora Suzen, P.G. Huang, S. Völker, A correlation-based transition model using local variables: Part i - model formulation. J. Turbomachinery (2004)
17. F.R. Menter, R.B. Langtry, Transition modelling for turbomachinery flows, in *Low Reynolds Number*, chapter 2, ed. by M.S. Genc (IntechOpen, Rijeka, 2012)
18. F.R. Menter, P.E.P. Smirnov, T. Liu, R. Avancha, A one-equation local correlation-based transition model. Flow, Turbulence Combust. **95**, 583–619 (2015)
19. S.C. Cakmakcioglu, O. Bas, R. Mura, U. Kaynak, A revised one-equation transitional model for external aerodynamics, in *AIAA Aviation 2020 Forum* (2020)
20. A.M. Savill, Further progress in the turbulence modelling of by-pass transition, in *Engineering Turbulence Modelling and Experiments*. ed. by W. Rodi, F. Martelli (Elsevier, Oxford, 1993), pp.583–592
21. R.J. McGhee, *Experimental Results for the Eppler 387 Airfoil at Low Reynolds Numbers in the Langley Low-Turbulence Pressure Tunnel*, vol. 4062 (National Aeronautics and Space Administration, Scientific and Technical Reports Server, 1988)
22. D.M. Somers, *Design and Experimental Results for the s809 Airfoil* (Airfoils Inc., State College, PA, 1989)
23. D.M. Somers, Design and experimental results for the s809 airfoil. Technical report, National Renewable Energy Lab. (NREL), Golden, CO (United States) (1997)
24. Y. Hoarau, D. Pena, J.B. Vos, D. Charbonnier, A. Gehri, M. Braza, T. Deloze, E. Laurendeau, Recent Developments of the Navier Stokes Multi Block (NSMB) CFD Solver (2016)
25. M.-S. Liou, C.J. Steffen, A new flux splitting scheme. J. Comput. Phys. **107**(1), 23–39 (1993)
26. B. van Leer, Towards the ultimate conservative difference scheme. v. a second-order sequel to godunov's method. J. Comput. Phys. **32**(1), 101–136 (1979)
27. R. Bourguet, M. Braza, G. Harran, R. El Akoury, Anisotropic organised eddy simulation for the prediction of non-equilibrium turbulent flows around bodies. J. Fluids Struct. **24**(8), 1240–1251 (2008). Unsteady separated flows and their control
28. A. Marouf, N. Simiriotis, J.B. Tô, Y. Bmegaptche, Y. Hoarau, M. Braza, Ddes and oes simulations of a morphing airbus a320 wing and flap in different scales at high reynolds, in *Progress in Hybrid RANS-LES Modelling*. ed. by Y. Hoarau, S.-H. Peng, D. Schwamborn, A. Revell, C. Mockett (Springer International Publishing, Cham, 2020), pp.249–258
29. A. Marouf. *Physical analysis of electroactive morphing concepts for the aerodynamic performance increase of future wing design through high-fidelity numerical simulation and turbulence modelling in high Reynols number*. PhD thesis, 2020. Thèse de doctorat dirigée par Hoarau, Yannick et Braza, Marianna Mécanique des fluides Strasbourg (2020)
30. A. Marouf, Y. Hoarau, J.-F. Rouchon, M. Braza, Three-dimensional simulation effects of trailing-edge actuation on a morphing a320 wing by means of hybrid turbulence modelling. Int. J. Numer. Meth. Heat Fluid Flow **33**, 1436–1457 (2023)

31. A. Marouf, A. Chouippe, J.B. Vos, D. Charbonnier, A. Gehri, M. Braza, Y. Hoarau, Unsteady cfd simulations for active flow control, in *AIAA Aviation 2021 Forum* (2021)
32. H.D. Truong, A. Marouf, Y. Hoarau, J.B. Vos, A. Gehri, Optimization of synthetic jet to enhance the aerodynamic performance of a vtol aircraft at real flight conditions, in *AIAA SCITECH 2023 Forum*
33. R. Langtry, F. Menter, Transition modeling for general CFD applications in aeronautics, in *43rd AIAA Aerospace Sciences Meeting and Exhibit* (American Institute of Aeronautics and Astronautics, 2005)
34. S. Medida, *Correlation-based transition modeling for external aerodynamic flows*. PhD thesis (University of Maryland, College Park, 2014)
35. A.J. Chorin, A numerical method for solving incompressible viscous flow problems. J. Comput. Phys. **2**(1), 12–26 (1967)

Exact Riemann Solution for a Homogeneous Equilibrium Model with Four Equations

Abdelmjid Qadi El Idrissi

Abstract This work presents a mathematical analysis to solve the Riemann problem for a two-phase gas–liquid mixture. For this purpose, we use the four-equation homogeneous equilibrium model with the Stiffened equation states. The exact solution of the Riemann problem is obtained by providing an explicit formulation of the different elementary waves: rarefaction, shock, and contact discontinuity. The validation of the proposed mathematical approach is confirmed by the good agreement found between the analytical solutions and the calculated results obtained by the numerical schemes of Mac Cormack and Godunov for various problems such as the Riemann problem, the partial Riemann problem for the treatment of boundary conditions, and the water hammer problem caused by a rapid closing valve.

1 Introduction

Two-phase gas–liquid flow continues to be an attractive topic for many researchers because it can occur in many natural and industrial applications such as petroleum, nuclear, hydraulics, and chemical engineering. In general, it is difficult to model a two-phase flow because the distribution of the gas inside the liquid can take on complex configurations. Therefore, the prediction of these flows becomes more complex and does not lead to a satisfactory approximation of the physical parameters of two-phase flows.

This study focuses on the case of a homogeneous mixture consisting of gas bubbles dispersed in the liquid. In the literature, several physical models have been developed to study this type of two-phase flow. According to the assumptions made on the thermodynamic equilibrium of the phases, the physical model can include a number ranging from three to seven equations. In the present work, we assume that the size of the dispersed particles is small enough to neglect the relative motion between phases.

A. Qadi El Idrissi (✉)
Hassan 1st University, National School of Applied Sciences, Berrechid, Morocco
e-mail: abdelmjid.qadi-elidrissi@uhp.ac.ma

D. Zeidan et al. (eds.), *Computational Fluid Dynamics: Novel Numerical and Computational Approaches*, Infosys Science Foundation Series,
https://doi.org/10.1007/978-981-97-8152-2_6

This means that the phases are strongly coupled and move at the same velocity. It is also assumed that the phases are in thermal and mechanical equilibrium, which implies that pressure and temperature are the same for liquid and gas. To model the evolution of the two-phase flow above, we use the four-equation homogeneous equilibrium model (HEM). It is a hyperbolic system of equations composed of two laws of conservation of mass for liquid and gas and two laws of conservation of momentum and energy for the mixture. To close the system, we use the Stiffened equations of state for the pure phases, which allow us to link the pressure and the temperature to the internal energy and the density, see [16–18, 20, 22].

In previous research, the HEM model has been widely used to predict different problems in fluid mechanics. However, few mathematical studies have been devoted to this model due to the strong non-linearity of its equations.

This work aims to develop a theoretical analysis of the HEM model by solving the Riemann problem. The analytical solution of this problem can provide valuable information on the physical behavior of the fluid mixture, especially when the flow generates the propagation of rarefaction or shock waves. This can help to understand the correlation that governs the different physical parameters such as pressure, void fraction, temperature, and density.

From a numerical point of view, the exact solution of the Riemann problem can help to build robust and efficient numerical schemes such as the finite volume methods: Godunov, WAF, etc. These schemes are known for their performance in ensuring the convergence of computed results but also for eliminating the numerical oscillations generated by the presence of shock waves due to the propagation of pressure waves, see [1, 2, 5, 9–12, 14, 19, 23–25, 27, 28].

In the theory of conservation laws, the Riemann problem is a hyperbolic system of equations whose initial state is a simple discontinuity composed of two states. It is demonstrated that the exact solution of this problem is defined as a combination of intermediate states separated by elementary waves: rarefaction, shock, and contact discontinuity, see [3–6, 21, 28].

In general, it is difficult to obtain an explicit form of these elementary waves due to the complexity of the equations. Despite the strong non-linearity of the four-equation homogeneous equilibrium model, a robust mathematical analysis has been developed, providing an explicit formulation of the different waves composing the Riemann solution.

To validate the robustness of the proposed mathematical approach, a series of test cases were carried out to verify the accuracy and performance of the analytical solution of the Riemann problem. To evaluate the theoretical calculation of the different waves, a comparison was made with the numerical results of the MacCormack scheme for different Riemann problems including rarefaction, contact discontinuity, and shock.

The treatment of boundary conditions was another interesting application using the notion of partial Riemann solutions. This can be used to solve the problem of water hammer generated by the sudden closing of the valve. Three case studies were considered: (a) propagation of a rarefaction wave, (b) propagation of a shock wave, and (c) propagation of a pressure wave between a closed valve and a tank with constant pressure, see [1, 5, 7, 8, 13, 15, 26]. The most interesting test remains the construction of a finite volume method implementing an exact Riemann solver. For all these tests, the validation was excellent, which confirms the robustness and efficiency of the mathematical approach.

A perspective of the present work is to derive an exact Riemann solver for the construction of robust finite volume methods such as Godunov, WAF, and TVD-WAF. Such finite volume methods are very accurate and effective in simulating complex physical problems such as water hammer, condensation, and evaporation problems. Moreover, these numerical methods are known for the convergence of their calculated results as well as for their ability to capture the shock wave.

The article is structured as follows: Sect. 2 presents the equations governing the four-equation homogeneous equilibrium model. Sections 3–7 give the definition of the eigenvalues, the corresponding eigenvectors, the nature of the different characteristics, and the formulation of the Riemann invariants. Sections 8–11 describe the theoretical structure of the different k-waves: shock, rarefaction, and contact discontinuity. Section 12 provides the formulation of the Riemann problem and the resolution procedure to obtain its analytical solution. Section 13 is devoted to the validation of the proposed mathematical analysis by comparing the numerical and exact results for various situations: Riemann problem, boundary condition, and water hammer problem. The conclusions are finally presented in Sect. 14.

2 Governing Equations

To describe the flow of the gas–liquid mixture, we use the HEM four-equation model. We assume that the two phases are

- in homogeneous flow model, i.e., have equal velocities,
- in thermal equilibrium, i.e., have equal temperatures,
- in mechanical equilibrium, i.e., have equal pressures.

This allows to formulate the HEM model by using two mass balance equations, and conservation equations for the total momentum and energy, see [16–18, 20].

Conservation of mass

$$\frac{\partial(\rho_g \alpha)}{\partial t} + \frac{\partial(\rho_g \alpha u)}{\partial x} = 0 \tag{1}$$

$$\frac{\partial(\rho_l (1-\alpha))}{\partial t} + \frac{\partial(\rho_l (1-\alpha) u)}{\partial x} = 0 \tag{2}$$

Conservation of momentum for the mixture

$$\frac{\partial(\rho u)}{\partial t} + \frac{\partial(\rho u^2 + p)}{\partial x} = 0 \tag{3}$$

where p is the pressure of the fluids.

Conservation of energy for the mixture

$$\frac{\partial(E)}{\partial t} + \frac{\partial(Eu + pu)}{\partial x} = 0 \tag{4}$$

where E denotes the mixture total energy.

To have a complete description of the fluid mixture, it is necessary to add the following constitutive laws:

Density of Mixture

$$\rho = \rho_g \alpha + \rho_l (1 - \alpha) \tag{5}$$

Energy of Mixture

$$E = E_g + E_l \tag{6}$$

$$E_k = \rho_k \alpha_k (e_k + \frac{1}{2} u^2), \quad k = g, l$$

where e_k is the internal energy.

Stiffened Equations of State

In the present study, we use the Stiffened equations of state EOS for the pure phases, see [16–18, 20]. With the Stiffened equations of state, we can link the pressure and the temperature to the internal energy and the density according to the following conditions:

$$p = (\gamma_k - 1)\rho_k (e_k - q_k) - \gamma_k p_\infty^k, \qquad k = g, l \tag{7}$$

$$\rho_k(p, T) = \frac{p + p_\infty^k}{(\gamma_k - 1) C_{vk} T}, \qquad k = g, l \tag{8}$$

where γ_k, q_k, p_∞^k, C_{vk}, and T are, respectively, the heat capacity ratio, the energy of the fluid k at a given reference state, a constant reference pressure, the thermal capacity of the fluid k, and the temperature of the fluids.

For the numerical analysis, we use the parameters of Stiffened equations given in Table 1.

From Eq. (7), we have $\rho_k e_k = q_k \rho_k + f_k(p)$
where

$$f_k(p) = \frac{1}{\gamma_k - 1}(p + \gamma_k p_\infty^k), \qquad k = g, l \tag{9}$$

This implies that: $E_k = q_k \rho_k \alpha_k + \alpha_k f_k(p) + \frac{1}{2}\rho_k \alpha_k u^2, \quad k = g, l$

Then, we have

$$\frac{\partial(E)}{\partial t} + \frac{\partial(Eu)}{\partial x} = \frac{\partial(\bar{E} + \frac{1}{2}\rho u^2)}{\partial t} + \frac{\partial((\bar{E} + \frac{1}{2}\rho u^2)u)}{\partial x} + \tag{10}$$

$$\sum_{k=g,l} q_k \left(\frac{\partial(\rho_k \alpha_k)}{\partial t} + \frac{\partial(\rho_k \alpha_k u)}{\partial x} \right)$$

where

$$\bar{E} = \bar{E}(\alpha, p) = \alpha f_g(p) + (1 - \alpha) f_l(p) \tag{11}$$

Since $\dfrac{\partial(\rho_k \alpha_k)}{\partial t} + \dfrac{\partial(\rho_k \alpha_k u)}{\partial x} = 0$, (k=g,l), we deduce that Eq. (4) is equivalent to:

$$\frac{\partial G}{\partial t} + \frac{\partial(Gu + pu)}{\partial x} = 0 \tag{12}$$

where

$$G = \bar{E} + \frac{1}{2}\rho u^2$$

Now, we can express Eqs. (1)–(4) under the conservative form:

$$\frac{\partial \mathcal{W}}{\partial t} + \frac{\partial F(\mathcal{W})}{\partial x} = 0 \tag{13}$$

where

$$\mathcal{W} = (\rho_g \alpha, \rho_l (1 - \alpha), \rho u, G)^T \tag{14}$$

$$F(\mathcal{W}) = (\rho_g \alpha u, \rho_l (1 - \alpha) u, \rho u^2 + p, (G + p)u)^T \tag{15}$$

2.1 Passage Between Primitive and Conservative Variables

As can be seen in (14), the conserved quantities $\mathcal{W}$ in terms of primitive variables $\mathcal{U} = (\alpha, p, u, T)^T$ are defined as follows:

$$\begin{cases} W_1 = \rho_g \alpha \\ W_2 = \rho_l (1-\alpha) \\ W_3 = (\rho_g \alpha + \rho_l (1-\alpha)) u \\ W_4 = f_g(p)\alpha + f_l(p)(1-\alpha) + \frac{1}{2}\rho u^2 \end{cases} \tag{16}$$

To obtain the primitive variables, we introduce the functions defined by

$$\bar{\rho}_k(p) = \frac{p + p_\infty^k}{(\gamma_k - 1)C_{vk}}, k = g, l \quad \bar{\rho}_m(\alpha, p) = \alpha \bar{\rho}_g(p) + (1-\alpha)\bar{\rho}_l(p) \tag{17}$$

This leads to the following resolution:

$$\begin{cases} u = \dfrac{W_3}{W_1 + W_2} \\ \dfrac{W_1}{\bar{\rho}_g(p)} + \dfrac{W_2}{\bar{\rho}_l(p)} = \dfrac{W_1 f_g(p)/\bar{\rho}_g(p) + W_2 f_l(p)/\bar{\rho}_l(p)}{W_4 - (W_1 + W_2)u^2/2} \\ T = \dfrac{1}{\dfrac{W_1}{\bar{\rho}_g(p)} + \dfrac{W_2}{\bar{\rho}_l(p)}} \\ \alpha = \dfrac{W_1 T}{\bar{\rho}_g(p)} \end{cases} \tag{18}$$

3 Quasi-Linear System

Using the previous primitive variables, the quasi-linear system of Equations (1)–(4) can be computed as

$$A(\mathcal{U})\frac{\partial \mathcal{U}}{\partial t} + B(\mathcal{U})\frac{\partial \mathcal{U}}{\partial x} = 0, \quad \mathcal{U} = \begin{pmatrix} \alpha \\ p \\ u \\ T \end{pmatrix} \tag{19}$$

Here, the matrices $A(\mathcal{U})$ and $B(\mathcal{U})$ are given by

$$A(\mathcal{U}) = \begin{pmatrix} 1 & \alpha E_g & 0 & \alpha K_g \\ -1 & (1-\alpha)E_l & 0 & \alpha K_l \\ 0 & 0 & \rho & 0 \\ a_p & a_\alpha & 0 & 0 \end{pmatrix} \quad B(\mathcal{U}) = \begin{pmatrix} u & \alpha E_g u & \alpha & \alpha K_g u \\ -u & (1-\alpha)E_l u & (1-\alpha) & (1-\alpha)K_l u \\ 0 & 1 & \rho u & 0 \\ a_p u & a_\alpha u & \bar{E}+p & 0 \end{pmatrix}$$

where

$$E_k = \frac{1}{\rho_k}\frac{\partial \rho_k}{\partial p} = \frac{1}{p+p_\infty^k}, \quad K_k = \frac{1}{\rho_k}\frac{\partial \rho_k}{\partial T} = -\frac{1}{T}, \quad k = g, l \tag{20}$$

and

$$a_p = f_g(p) - f_l(p) = \frac{\gamma_g(p+p_\infty^g)}{\gamma_g - 1} - \frac{\gamma_l(p+p_\infty^l)}{\gamma_l - 1} = \frac{\gamma_g}{\gamma_g - 1}\frac{1}{E_g} - \frac{\gamma_l}{\gamma_l - 1}\frac{1}{E_l} \tag{21}$$

$$a_\alpha = \frac{\alpha}{\gamma_g - 1} + \frac{1-\alpha}{\gamma_l - 1} \tag{22}$$

4 Eigenvalues

The eigenvalues λ are defined to be solutions of the equation:

$$det\Big(B(\mathcal{U}) - \lambda A(\mathcal{U})\Big) = 0$$

where

$$det\Big(B(\mathcal{U}) - \lambda A(\mathcal{U})\Big) = (u-\lambda)^2 \begin{vmatrix} 1 & \alpha E_g(u-\lambda) & \alpha & \alpha K_g \\ -1 & (1-\alpha)E_l(u-\lambda) & (1-\alpha) & (1-\alpha)K_l \\ 0 & 1 & \rho(u-\lambda) & 0 \\ a_p & a_\alpha(u-\lambda) & \bar{E}+p & 0 \end{vmatrix}$$

By developing this determinant, we obtain

$$det\Big(B(\mathcal{U}) - \lambda A(\mathcal{U})\Big) = \frac{1}{T}(u-\lambda)^2\Big[\rho(u-\lambda)^2\Big(a_\alpha - a_p\alpha(1-\alpha)(E_g - E_l)\Big) - \bar{E} - p\Big]$$

Under the hypothesis: $\gamma_k > 1, k = g, l$, we prove in the lemma below that the system (1)–(4) is hyperbolic and has the following eigenvalues:

$$\begin{cases} \lambda_1 = u - c \\ \lambda_{2,3} = u \\ \lambda_4 = u + c \end{cases} \tag{23}$$

where c is the speed of sound in the fluid mixture:

$$c = \frac{1}{\sqrt{\rho K}} \tag{24}$$

and

$$K = K(\alpha, p) = \frac{a_\alpha - a_p \alpha(1-\alpha)(E_g - E_l)}{\bar{E} + p} \tag{25}$$

In order to analyze the nature of the k-characteristic fields and establish the k-waves (rarefaction, shock), it is convenient to rewrite the expression of $K(\alpha, p)$ as a function of the pressure p and the variable $\mathcal{J}$:

$$\mathcal{J} = \frac{\alpha(p + p_\infty^g)}{(1-\alpha)(p + p_\infty^l)} \tag{26}$$

Lemma 4.1

Under the condition $\gamma_k > 1$, $(k = g, l)$, the system (1)–(4) is hyperbolic and admits the following real eigenvalues:

$$\begin{cases} \lambda_1 = u - c \\ \lambda_{2,3} = u \\ \lambda_4 = u + c \end{cases} \tag{27}$$

where $c = \dfrac{1}{\sqrt{\rho K}}$, and K is a rational function defined by the following expression:

$$K = K(p, \mathcal{J}) = \frac{\mathcal{J}}{A_{g,l}} \frac{R(p, \mathcal{J})}{Q(p, \mathcal{J})} \tag{28}$$

$R(p, \mathcal{J})$ and $Q(p, \mathcal{J})$ are two polynomial functions given by

$$R(p, \mathcal{J}) = C_2(p + p_\infty^g)^2 + C_1(p + p_\infty^g) + C_0 \tag{29}$$

$$Q(p, \mathcal{J}) = (p + p_\infty^g)(p + p_\infty^l)\Big((1 + \mathcal{J})p + p_\infty^g + \mathcal{J} p_\infty^l\Big) \tag{30}$$

and

$$A_{gl} = \left(\frac{\gamma_g \mathcal{J}}{\gamma_g - 1} + \frac{\gamma_l}{\gamma_l - 1}\right) \quad \bar{A}_{g,l} = \frac{\mathcal{J}}{\gamma_g - 1} + \frac{1}{\gamma_l - 1} \tag{31}$$

$$C_2 = \frac{1 + \mathcal{J}}{\mathcal{J}} \bar{A}_{g,l} \quad C_1 = 2\bar{A}_{g,l}(p_\infty^l - p_\infty^g) \quad C_0 = (\bar{A}_{g,l} + 1)(p_\infty^l - p_\infty^g)^2 \tag{32}$$

Moreover, the function $K(p, \mathcal{J})$ verifies the following property:

$$\frac{\partial K(p, \mathcal{J})}{\partial p} < 0 \tag{33}$$

Proof From the definition (26) of $\mathcal{J}$, we can express the void fraction as follows:

$$\alpha = \frac{\mathcal{J} E_g}{E_l + \mathcal{J} E_g}, \qquad 1 - \alpha = \frac{E_l}{E_l + \mathcal{J} E_g} \tag{34}$$

To get the result (28), we have to substitute the void fraction (34) into the expression (25). Following Eq. (9), we have

$$f_k(p) + p = \frac{1}{\gamma_k - 1}(p + \gamma_k p_\infty^k) + p = \frac{\gamma_k}{\gamma_k - 1}(p + p_\infty^k) \qquad k = g, l \tag{35}$$

It follows that

$$\bar{E} + p = \alpha(f_g(p) + p) + (1 - \alpha)(f_l(p) + p) = \frac{\gamma_g}{\gamma_g - 1}\frac{\alpha}{E_g} + \frac{\gamma_l}{\gamma_l - 1}\frac{1 - \alpha}{E_l} \tag{36}$$

We deduce that:

$$\bar{E} + p = \frac{A_{gl}}{E_l + \mathcal{J} E_g}, \; where \quad A_{gl} = \Big(\frac{\gamma_g \mathcal{J}}{\gamma_g - 1} + \frac{\gamma_l}{\gamma_l - 1}\Big) \tag{37}$$

On the other hand, we have

$$a_\alpha - a_p \alpha(1 - \alpha)(E_g - E_l) = \alpha(1 - \alpha) E_g E_l \Big(\frac{a_\alpha}{\alpha(1 - \alpha) E_g E_l} - a_p(\frac{1}{E_l} - \frac{1}{E_g})\Big) \tag{38}$$

According to Eq. (34), we have

$$\alpha(1 - \alpha) E_g E_l = \mathcal{J}\Big(\frac{E_g E_l}{E_l + \mathcal{J} E_g}\Big)^2 = \frac{\mathcal{J}}{(E_l + \mathcal{J} E_g) Q(p)} \tag{39}$$

where

$$Q(p, \mathcal{J}) = \frac{1}{E_g}\frac{1}{E_l}(\frac{1}{E_g} + \frac{\mathcal{J}}{E_l}) \tag{40}$$

By using Eq. (22), we can also write

$$\frac{a_\alpha}{\alpha(1 - \alpha) E_g E_l} = \frac{1}{E_g E_l}\Big(\frac{1}{\gamma_g - 1}\frac{1}{1 - \alpha} + \frac{1}{\gamma_l - 1}\frac{1}{\alpha}\Big) = \Big(\frac{1}{\gamma_g - 1}\frac{1}{E_l} + \frac{1}{\gamma_l - 1}\frac{1}{\mathcal{J} E_g}\Big)(\frac{1}{E_g} + \frac{\mathcal{J}}{E_l}) \tag{41}$$

Let's introduce $R(p,\mathcal{J})$ the function defined by

$$R(p,\mathcal{J}) = \frac{a_\alpha}{\alpha(1-\alpha)E_g E_l} - a_p(\frac{1}{E_l} - \frac{1}{E_g}) \tag{42}$$

Then, we have

$$R(p,\mathcal{J}) = \frac{A_{g,l}-1}{\mathcal{J}E_g^2} + \frac{A_{g,l}-\mathcal{J}}{E_l^2} - \frac{2}{E_g E_l} \tag{43}$$

It follows that:

$$a_\alpha - a_p\alpha(1-\alpha)(E_g - E_l) = \frac{\mathcal{J}}{E_l + \mathcal{J}E_g}\frac{R(p,\mathcal{J})}{Q(p,\mathcal{J})} \tag{44}$$

We conclude that

$$K(p,\mathcal{J}) = \frac{\mathcal{J}}{A_{g,l}}\frac{R(p,\mathcal{J})}{Q(p,\mathcal{J})} \tag{45}$$

To demonstrate that the system (1)–(4) is hyperbolic, it suffices to verify that the function $R(p,\mathcal{J})$ is positive. □

To achieve this, the function $R(p,\mathcal{J})$ needs to be written as a polynomial of second degree. By using Eq. (20), we can write

$$R(p,\mathcal{J}) = C_2(p+p_\infty^g)^2 + C_1(p+p_\infty^g) + C_0 \tag{46}$$

where

$$C_2 = \frac{1+\mathcal{J}}{\mathcal{J}}\bar{A}_{g,l} \quad C_1 = 2\bar{A}_{g,l}(p_\infty^l - p_\infty^g) \quad C_0 = (\bar{A}_{g,l}+1)(p_\infty^l - p_\infty^g)^2 \tag{47}$$

and

$$\bar{A}_{g,l} = \frac{\mathcal{J}}{\gamma_l - 1} + \frac{1}{\gamma_l - 1} \tag{48}$$

This polynomial admits a delta given by

$$\Delta = C_1^2 - 4C_0C_2 = -4\bar{A}_{g,l}(1+\mathcal{J}+\bar{A}_{g,l})(p_\infty^l - p_\infty^g)^2 \tag{49}$$

Since $\Delta < 0$ *and* $C_2 > 0$, we deduce that $R(p,\mathcal{J}) > 0$.

This means that the system (1)–(4) is hyperbolic with real eigenvalues.

Now, the function $K(p,\mathcal{J})$ can be written as follows

$$K(p,\mathcal{J}) = \frac{\mathcal{J}}{A_{g,l}}\frac{C_2(p+p_\infty^g)^2 + C_1(p+p_\infty^g) + C_0}{(p+p_\infty^g)(p+p_\infty^l)\Big(p+p_\infty^g+\mathcal{J}(p+p_\infty^l)\Big)} \tag{50}$$

By developing this rational function, we obtain the following decomposition:

$$K(p,\mathcal{J}) = \frac{a^l}{p+p_\infty^l} + \frac{a^g}{p+p_\infty^g} - \frac{1}{p+p_\infty^g+a^J} \tag{51}$$

where

$$a^g = 1-\mathcal{J}/A_{g,l} \quad a^l = 1-1/A_{g,l} \quad a^J = \frac{\mathcal{J}}{1+\mathcal{J}}(p_\infty^l - p_\infty^g) \tag{52}$$

It is easy to see that

$$\frac{\partial K(p,\mathcal{J})}{\partial p} < 0 \tag{53}$$

To demonstrate this, we have

$$\frac{\partial K(p,\mathcal{J})}{\partial p} = -\frac{a^l}{(p+p_\infty^l)^2} - \frac{a^g}{(p+p_\infty^g)^2} + \frac{1}{(p+p_\infty^g+a^J)^2} \tag{54}$$

Then, we get

$$\frac{\partial K(p,\mathcal{J})}{\partial p} = -\frac{a^l}{(p+p_\infty^l)^2} + \frac{\mathcal{J}}{A_{g,l}} \frac{\left(p+p_\infty^g-r_1\right)\left(p+p_\infty^g-r_2\right)}{(p+p_\infty^g)^2(p+p_\infty^g+a^J)^2} \tag{55}$$

where

$$\begin{cases} r_1 = -\dfrac{a^J\sqrt{a^g}}{\sqrt{a^g}+1} \\ r_2 = \dfrac{a^J\sqrt{a^g}}{1-\sqrt{a^g}} = \dfrac{a^J}{\mathcal{J}} A_{g,l}\sqrt{a^g}(1+\sqrt{a^g}) \end{cases} \tag{56}$$

Since $\mathcal{J} \approx 0$, we have then $a^g \approx 1$

$$\begin{cases} r_1 = -\dfrac{a^J}{2} = -\dfrac{\mathcal{J}}{2(1+\mathcal{J})}(p_\infty^l - p_\infty^g) \le 0 \\ r_2 = \dfrac{a^J\sqrt{a^g}}{1-\sqrt{a^g}} = \dfrac{2A_{g,l}}{1+\mathcal{J}}(p_\infty^l - p_\infty^g) \approx 2(p_\infty^l - p_\infty^g) \end{cases} \tag{57}$$

Here the inequalities $r_1 \le p + p_\infty^g \le r_2$ hold for all the pressure, we deduce that

$$\left(p+p_\infty^g-r_1\right)\left(p+p_\infty^g-r_2\right) \le 0 \tag{58}$$

Consequently

$$\frac{\partial K(p,\mathcal{J})}{\partial p} < 0 \tag{59}$$

5 Right Eigenvector

For each eigenvalue λ_k, we denote by $r_k(\mathcal{U}) = (X_1, X_2, X_3, X_4)^T$ the right eigenvector defined as solution of the linear system:

$$\Big(B(\mathcal{U}) - \lambda_k A(\mathcal{U})\Big).r_k(\mathcal{U}) = 0 \quad , \quad k = 1, 2, 3, 4 \tag{60}$$

Using the expressions of the matrices $A(\mathcal{U})$ and $B(\mathcal{U})$, Eq. (60) becomes

$$\begin{cases} (u - \lambda_k)X_1 + \alpha E_g(u - \lambda_k)X_2 + \alpha X_3 + \alpha K_g(u - \lambda_k)X_4 = 0 \\ -(u - \lambda_k)X_1 + (1 - \alpha)E_l(u - \lambda_k)X_2 + (1 - \alpha)X_3 + (1 - \alpha)K_l(u - \lambda_k)X_4 = 0 \\ X_2 + \rho(u - \lambda_k)X_3 = 0 \\ a_p(u - \lambda_k)X_1 + a_\alpha(u - \lambda_k)X_2 + (\bar{E} + p)X_3 = 0 \end{cases} \tag{61}$$

The resolution of this system leads to the following right eigenvectors:

a. Case $\lambda_k = u$

$$r_2(\mathcal{U}) = \begin{pmatrix} 1 \\ 0 \\ 0 \\ 0 \end{pmatrix} \quad , \quad r_3(\mathcal{U}) = \begin{pmatrix} 0 \\ 0 \\ 0 \\ 1 \end{pmatrix} \tag{62}$$

b. Case $\lambda_k \neq u$

$$r_{1/4}(\mathcal{U}) = \begin{pmatrix} -\alpha(1 - \alpha)(E_g - E_l) \\ 1 \\ \dfrac{1}{\rho(\lambda_k - u)} \\ -T\left(\dfrac{1}{\rho(\lambda_k - u)^2} - E_m\right) \end{pmatrix} \tag{63}$$

where $E_m = \alpha E_g + (1 - \alpha)E_l$

6 Linear Degeneracy and Genuinely Non-Linear

By definition, an k-characteristic field is called linearly degenerate when $\nabla\lambda_k.r_k(\mathcal{U}) = 0$. It is genuinely non-linear if $\nabla\lambda_k.r_k(\mathcal{U}) \neq 0$

Lemma 6.1

For the system (1)–(4), we have the following proprieties:

- The second and third characteristic fields are linearly degenerate.
- The first and fourth characteristic fields are genuinely non-linear.

Furthermore, we have the properties:

$$\nabla\lambda_1.r_1(\mathcal{U}) < 0 \quad and \quad \nabla\lambda_4.r_4(\mathcal{U}) > 0 \tag{64}$$

Proof For the characteristic fields $\lambda_{2/3} = u$, the Laplacian is computed as $\nabla\lambda_{2/3} = (0, 0, 1, 0)^T$.

Then, we have

$$\nabla\lambda_2.r_2(\mathcal{U}) = (0, 0, 1, 0)^T.(1, 0, 0, 0)^T = 0$$

$$\nabla\lambda_3.r_3(\mathcal{U}) = (0, 0, 1, 0)^T.(0, 0, 0, 1)^T = 0$$

We therefore deduce that $\lambda_{2/3}$ is linearly degenerate. □

For $\lambda_k = u \pm c$, we determine the Laplacian as

$$\nabla\lambda_k = \Big(\pm\frac{\partial c}{\partial\alpha}, \pm\frac{\partial c}{\partial p}, 1, \pm\frac{\partial c}{\partial T}\Big)^T \tag{65}$$

It follows that

$$\nabla\lambda_k.r_k(\mathcal{U}) = (\lambda_k - u)\Big(K + \frac{1}{c}\nabla c.r_k(\mathcal{U})\Big) \tag{66}$$

Using Eq. (24), the above equation can be developed as

$$K + \frac{1}{c}\nabla c.r_k(\mathcal{U}) = \frac{1}{2}\Big(2K - \frac{1}{\rho}\nabla\rho.r_k(\mathcal{U}) - \frac{1}{K}\nabla K.r_k(\mathcal{U})\Big) \tag{67}$$

From Eq. (5), the Laplacian of the mixture density can be computed as

$$\nabla\rho = \Big(\rho_g - \rho_l, \rho_g\alpha E_g + \rho_l(1-\alpha)E_l, 0, -\frac{\rho}{T}\Big)^T$$

Then, we deduce that

$$\nabla \rho . r_k(\mathcal{U}) = -\alpha(1-\alpha)(E_g - E_l)(\rho_g - \rho_l) + \rho_g \alpha E_g + \rho_l(1-\alpha)E_l + \rho(K - E_m)$$

which leads to

$$\nabla \rho . r_k(\mathcal{U}) = \rho K \tag{68}$$

On the other hand, the use of Eq. (28) allows to compute the gradient of K as

$$\nabla K(p, \mathcal{J}) = \Big(\frac{\partial K}{\partial \mathcal{J}}\frac{\partial \mathcal{J}}{\partial \alpha}, \frac{\partial K}{\partial p} + \frac{\partial K}{\partial \mathcal{J}}\frac{\partial \mathcal{J}}{\partial p}, 0, 0\Big)^T$$

It follows that

$$\nabla K . r_k(\mathcal{U}) = \Big(-\alpha(1-\alpha)(E_g - E_l)\frac{\partial \mathcal{J}}{\partial \alpha} + \frac{\partial \mathcal{J}}{\partial p}\Big)\frac{\partial K}{\partial \mathcal{J}} + \frac{\partial K}{\partial p} \tag{69}$$

Since the function $\mathcal{J}$ verify the property:

$$-\alpha(1-\alpha)(E_g - E_l)\frac{\partial \mathcal{J}}{\partial \alpha} + \frac{\partial \mathcal{J}}{\partial p} = 0 \tag{70}$$

We therefore deduce that

$$\nabla K . r_k(\mathcal{U}) = \frac{\partial K}{\partial p} \leq 0 \tag{71}$$

Equations (68) and (71) demonstrate that

$$2K - \frac{1}{\rho}\nabla \rho . r_k(\mathcal{U}) - \frac{1}{K}\nabla K . r_k(\mathcal{U}) = \frac{1}{K}\Big(K^2 - \frac{1}{K}\frac{\partial K}{\partial p}\Big)$$

Thus, we have

$$K + \frac{1}{c}\nabla c . r_k(\mathcal{U}) > 0 \tag{72}$$

We conclude from Eq. (66) that the first and fourth characteristic fields are genuinely non-linear. Furthermore, we have the following proprieties:

$$\nabla \lambda_1 . r_1(\mathcal{U}) < 0, \quad and \quad \nabla \lambda_4 . r_4(\mathcal{U}) > 0 \tag{73}$$

7 k-Riemann Invariants

A smooth function $\mathcal{I}_k(\mathcal{U})$ is called an k-Riemann invariant when it satisfies

$$\nabla \mathcal{I}_k(\mathcal{U}) . r_k(\mathcal{U}) = 0, \quad k = 1, 2, 3, 4 \tag{74}$$

Lemma 7.1

The k-Riemann invariants associated with the system (1)–(4) are given by the following expressions:
1-Riemann invariants

$$\begin{cases} \mathcal{J}(\mathcal{U}) = \dfrac{\alpha(p+p_\infty^g)}{(1-\alpha)(p+p_\infty^l)} \\ \\ \mathcal{I}(\mathcal{U}) = \dfrac{T}{\phi(p,\mathcal{J})} \\ \\ \mathcal{K}_1(\mathcal{U}) = u + \psi(p,\mathcal{I},\mathcal{J}) \end{cases} \tag{75}$$

2/3-Riemann invariants

$$\mathcal{J}_{2,3}(\mathcal{U}) = p \quad , \quad \mathcal{I}_{2,3}(\mathcal{U}) = u \tag{76}$$

4-Riemann invariants

$$\begin{cases} \mathcal{J}(\mathcal{U}) = \dfrac{\alpha(p+p_\infty^g)}{(1-\alpha)(p+p_\infty^l)} \\ \\ \mathcal{I}(\mathcal{U}) = \dfrac{T}{\phi(p,\mathcal{J})} \\ \\ \mathcal{K}_4(\mathcal{U}) = u - \psi(p,\mathcal{I},\mathcal{J}) \end{cases} \tag{77}$$

where

$$\phi(p,\mathcal{J}) = (p+p_\infty^g)^{\frac{\mathcal{J}}{\Lambda_{g,l}}}(p+p_\infty^l)^{\frac{1}{\Lambda_{g,l}}} \tag{78}$$

$$\psi(p,\mathcal{I},\mathcal{J}) = \int^p \frac{1}{\rho c}\,dp = \psi(p,\mathcal{I},\mathcal{J}) = \sqrt{\frac{\mathcal{J}}{C_m A_{g,l}}}\int^p \frac{\sqrt{R(p,\mathcal{J})}}{(p+p_\infty^g)^{n_1/2}(p+p_\infty^l)^{n_2/2}}\,dp \tag{79}$$

Proof To solve Eq. (74), we have to consider two cases:

a. Case $\lambda_k = u$

It is easy to verify that $\mathcal{I}_k(\mathcal{U}) = u$ *and* $\mathcal{J}_k(\mathcal{U}) = p$ are 2/3-Riemann Invariants.

To demonstrate this, we compute the Laplacian of these functions as

$$\nabla \mathcal{I}_k(\mathcal{U}) = (0, 0, 1, 0)^T \quad and \quad \nabla \mathcal{J}_k(\mathcal{U}) = (0, 1, 0, 0)^T \tag{80}$$

Using the expression (62) for the 2/3 right eigenvectors, we find that

$$\nabla \mathcal{I}_k(\mathcal{U}).r_2(\mathcal{U}) = \nabla \mathcal{I}_k(\mathcal{U}).r_3(\mathcal{U}) = 0, \quad and \quad \nabla \mathcal{J}_k(\mathcal{U}).r_2(\mathcal{U}) = \nabla \mathcal{J}_k(\mathcal{U}).r_3(\mathcal{U}) = 0 \tag{81}$$

This confirms that $\mathcal{J}_{2/3}(\mathcal{U})$ and $\mathcal{I}_{2/3}(\mathcal{U})$ are two 2/3-Riemann Invariants. □

b. Case $\lambda_k \neq u$

For a k-characteristic field genuinely non-linear, the practical way to determine the k-Riemann Invariant $\mathcal{I}_k(\mathcal{U})$ is to consider the following differential system:

$$\frac{d}{ds}(\mathcal{U}_k(s)) = r_k(\mathcal{U}_k(s)), \quad k = 1, 4 \tag{82}$$

Applying the definition (63) of the 1/4 right eigenvectors, Eq. (82) can be expressed as

$$\begin{cases} \dfrac{d\alpha}{ds} = -\alpha(1-\alpha)(E_g - E_l) \\ \dfrac{dp}{ds} = 1 \\ \dfrac{du}{ds} = \dfrac{1}{\rho(\lambda_k - u)} \\ \dfrac{dT}{ds} = -T\Big(\dfrac{1}{\rho(\lambda_k - u)^2} - E_m\Big) \end{cases} \tag{83}$$

It follows that

$$\begin{cases} \dfrac{d\alpha}{dp} = -\alpha(1-\alpha)\Big(\dfrac{1}{p + p_\infty^g} - \dfrac{1}{p + p_\infty^l}\Big) \\ \dfrac{dT}{dp} = -T\Big(\dfrac{1}{\rho(\lambda_k - u)^2} - E_m\Big) \\ \dfrac{du}{dp} = \dfrac{1}{\rho(\lambda_k - u)} = \pm\dfrac{1}{\rho c} \end{cases} \tag{84}$$

From the first differential, we can write

$$\frac{d\alpha}{\alpha(1-\alpha)} = -\Big(\frac{1}{p+p_\infty^g} - \frac{1}{p+p_\infty^l}\Big)dp \tag{85}$$

By integrating this differential equation, we obtain an expression of the 1/4-Riemann invariants in the following form:

$$\mathcal{J} = \mathcal{J}(\mathcal{U}) = \frac{\alpha(p+p_\infty^g)}{(1-\alpha)(p+p_\infty^l)} = \frac{\alpha E_l}{(1-\alpha)E_g} \tag{86}$$

This implies that

$$\alpha = \frac{\mathcal{J}E_g}{E_l+\mathcal{J}E_g}, \qquad 1-\alpha = \frac{E_l}{E_l+\mathcal{J}E_g} \tag{87}$$

To solve the second differential equation of (84), we assume that $\mathcal{J}(\mathcal{U})$ is constant by virtue of Eq. (86), which gives the following solution:

$$Log(T) = -\int^p \Big(\frac{1}{\rho(\lambda_k-u)^2} - E_m\Big)dp \tag{88}$$

In this integral, the term E_m should be expressed as a function of E_g and E_l. To achieve this, we use Eq. (87) to write

$$E_m = \alpha E_g + (1-\alpha)E_l = \frac{\mathcal{J}E_g^2 + E_l^2}{E_l+\mathcal{J}E_g} = \frac{(E_gE_l)^2}{E_l+\mathcal{J}E_g}\Big(\frac{\mathcal{J}}{E_l^2}+\frac{1}{E_g^2}\Big) \tag{89}$$

This means that

$$E_m = \frac{R'(p,\mathcal{J})}{Q(p,\mathcal{J})}, \; where \quad R'(p,\mathcal{J}) = \frac{\mathcal{J}}{E_l^2} + \frac{1}{E_g^2} \tag{90}$$

It follows that

$$\frac{1}{\rho(\lambda_k-u)^2} - E_m = \frac{1}{A_{g,l}}\frac{P(p,\mathcal{J})}{Q(p,\mathcal{J})} \tag{91}$$

where

$$P(p,\mathcal{J}) = \mathcal{J}R(p,\mathcal{J}) - A_{g,l}R'(p,\mathcal{J}) \tag{92}$$

By using the expressions of $R(p,\mathcal{J})$ and $R'(p,\mathcal{J})$, we demonstrate easily that

$$P(p,\mathcal{J}) = -\Big(\frac{1}{E_g} + \frac{\mathcal{J}}{E_l}\Big)^2 \tag{93}$$

Then, we have

$$\frac{1}{\rho(\lambda_k - u)^2} - E_m = -\frac{1}{A_{g,l}}\left(\frac{\mathcal{J}}{p + p_\infty^g} + \frac{1}{p + p_\infty^l}\right) \tag{94}$$

Substituting this in Eq. (88), we obtain a correlation between the temperature and the pressure under the form:

$$Log(T) = log\left((p + p_\infty^g)^{\frac{\mathcal{J}}{A_{g,l}}}(p + p_\infty^l)^{\frac{1}{A_{g,l}}}\right) + constant \tag{95}$$

Thus, an expression of the 1/4-Riemann Invariants can be derived as follows:

$$\mathcal{I}(\mathcal{U}) = \frac{T}{\phi(p, \mathcal{J})} \tag{96}$$

where

$$\phi(p, \mathcal{J}) = (p + p_\infty^g)^{\frac{\mathcal{J}}{A_{g,l}}}(p + p_\infty^l)^{\frac{1}{A_{g,l}}} \tag{97}$$

To solve the third differential equation of (84), we have to write

$$\frac{du}{dp} = \pm\frac{1}{\rho(\lambda_k - u)} = \pm\sqrt{\frac{K}{\rho}} \tag{98}$$

Using the relations (8) and (87), the mixture density can be rewritten as

$$\rho = \rho_g\alpha + \rho_l(1 - \alpha) = \rho_l(1 - \alpha)\left(1 + \bar{\mathcal{J}}\right) = \frac{1 + \bar{\mathcal{J}}}{(\gamma_l - 1)C_{vl}T(E_l + \mathcal{J}E_g)} \tag{99}$$

where

$$\bar{\mathcal{J}} = \mathcal{J}\frac{(\gamma_l - 1)C_{vl}}{(\gamma_g - 1)C_{vg}} \tag{100}$$

This yields to

$$\rho = C_m\frac{(p + p_\infty^g)^2(p + p_\infty^l)^2}{Q(p)\phi(p, \mathcal{J})} = C_m\frac{(p + p_\infty^g)^{n_1}(p + p_\infty^l)^{n_2}}{Q(p)} \tag{101}$$

where

$$C_m = \frac{1 + \bar{\mathcal{J}}}{(\gamma_l - 1)C_{vl}\mathcal{I}} \quad n_1 = 2 - \frac{\mathcal{J}}{A_{g,l}} \quad and \quad n_2 = 2 - \frac{1}{A_{g,l}} \tag{102}$$

Now, we can use Equations (101) and (45) to formulate the differential Equation (98) as follows:

$$\frac{du}{dp} = \pm\sqrt{\frac{\mathcal{J}}{C_m A_{g,l}}\frac{\sqrt{R(p, \mathcal{J})}}{(p + p_\infty^g)^{n_1/2}(p + p_\infty^l)^{n_2/2}}} \tag{103}$$

An integration of this equation yields to define the 1/4-Riemann Invariants according to:

$$\mathcal{K}_k(\mathcal{U}) = u \mp \psi(p, \mathcal{I}, \mathcal{J}) \tag{104}$$

where

$$\psi(p, \mathcal{I}, \mathcal{J}) = \sqrt{\frac{\mathcal{J}}{C_m A_{g,l}}} \int^p \frac{\sqrt{R(p, \mathcal{J})}}{(p + p^g_\infty)^{n_1/2}(p + p^l_\infty)^{n_2/2}} dp \tag{105}$$

and $R(p, \mathcal{J})$ is the polynomial of second degree defined by (29):

$$R(p, \mathcal{J}) = C_2(p + p^g_\infty)^2 + C_1(p + p^g_\infty) + C_0$$

To obtain an explicit expression of ψ, it is necessary to apply some physical considerations relative to the flow conditions of the gas–liquid mixture, which can greatly simplify the formula inside the integral.

First, the quantity $p + p^l_\infty$ is assumed to be constant due to the strong high value of $p + p^l_\infty$ compared to the pressure p.

Secondly, the value of $\mathcal{J}$ is so small that we can consider constant the exponent n_1, then we have $n_1 = 2$.

By doing this, we get an explicit expression for the integral (105) using the following function:

$$\begin{aligned}\psi(p, \mathcal{I}, \mathcal{J}) = \sqrt{\frac{\mathcal{J}}{C_m A_{g,l}}}(p + p^l_\infty)^{-n_2/2}\Big[\sqrt{R(p, \mathcal{J})} + \frac{C_1}{2\sqrt{C_2}} log\Big(\sqrt{R(p, \mathcal{J})} + \sqrt{C_2}(p + p^g_\infty) + \frac{C_1}{2\sqrt{C_2}}\Big) \\ - \sqrt{C_0} log\Big(\sqrt{R(p, \mathcal{J}} + \frac{C_1}{2\sqrt{C_0}}(p + p^g_\infty) + \sqrt{C_0}\Big) + \sqrt{C_0} log(p + p^g_\infty)\Big]\end{aligned} \tag{106}$$

Finally, the k-Riemann invariants can be defined as follows:

1-Riemann invariants

$$\begin{cases} \mathcal{J}(\mathcal{U}) = \dfrac{\alpha(p + p^g_\infty)}{(1 - \alpha)(p + p^l_\infty)} \\ \\ \mathcal{I}(\mathcal{U}) = \dfrac{T}{\phi(p, \mathcal{J})} \\ \\ \mathcal{K}_1(\mathcal{U}) = u + \psi(p, \mathcal{I}, \mathcal{J}) \end{cases} \tag{107}$$

2/3-Riemann invariants

$$\mathcal{J}_{2,3}(\mathcal{U}) = p \quad , \quad \mathcal{I}_{2,3}(\mathcal{U}) = u \tag{108}$$

4-Riemann invariants

$$\begin{cases} \mathcal{J}(\mathcal{U}) = \dfrac{\alpha(p + p_\infty^g)}{(1-\alpha)(p + p_\infty^l)} \\ \\ \mathcal{I}(\mathcal{U}) = \dfrac{T}{\phi(p, \mathcal{J})} \\ \\ \mathcal{K}_4(\mathcal{U}) = u - \psi(p, \mathcal{I}, \mathcal{J}) \end{cases} \tag{109}$$

8 k-Rarefaction Wave

Let λ_k be a genuinely non-linear characteristic. A smooth function $\mathcal{U}_k(\xi)$ is called a k-rarefaction wave $\mathcal{U}_k(\xi)$ when it connects a left state $\mathcal{U}_l = (\alpha_l, p_l, u_l, T_l)^T$ to a right state $\mathcal{U}_r = (\alpha_r, p_r, u_r, T_r)^T$ by verifying the following conditions:

$$\begin{cases} \dfrac{d}{d\xi}(\mathcal{U}_k(\xi)) = r_k(\mathcal{U}_k(\xi)) \\ \\ \lambda_k(\mathcal{U}_k(\xi)) = \xi \\ \\ \lambda_k(\mathcal{U}_l) \leq \xi \leq \lambda_k(\mathcal{U}_r) \end{cases} \tag{110}$$

where λ_k is a genuinely non-linear characteristic.

In this definition, it can be noted that the k-rarefaction wave is the solution of a system of differential equations coupled with a root problem.

Lemma 8.1

The k-rarefaction waves corresponding to the first and fourth characteristic fields are defined as follows:

1-Rarefaction Wave

$$\begin{cases} \alpha_r = \varphi(p_r, \mathcal{J}) \\ \\ T_r = T_l \dfrac{\phi(p_r, \mathcal{J})}{\phi(p_l, \mathcal{J})} \\ \\ u_r = u_l - \Big(\psi(p_r, \mathcal{I}, \mathcal{J}) - \psi(p_l, \mathcal{I}, \mathcal{J})\Big) \qquad p_r \leq p_l \end{cases} \tag{111}$$

4-Rarefaction Wave

$$\begin{cases} \alpha_r = \varphi(p_r, \mathcal{J}) \\ T_r = T_l \dfrac{\phi(p_r, \mathcal{J})}{\phi(p_l, \mathcal{J})} \\ u_r = u_l + \Big(\psi(p_r, \mathcal{I}, \mathcal{J}) - \psi(p_l, \mathcal{I}, \mathcal{J})\Big) \qquad p_l \leq p_r \end{cases} \tag{112}$$

where

$$\mathcal{I} = \mathcal{I}(\mathcal{U}_l) = \mathcal{I}(\mathcal{U}_r), \qquad \mathcal{J} = \mathcal{J}(\mathcal{U}_l) = \mathcal{J}(\mathcal{U}_r) \tag{113}$$

$$\varphi(p, \mathcal{J}) = \frac{\mathcal{J}(p + p^l_\infty)}{p + p^g_\infty + \mathcal{J}(p + p^l_\infty)}, \qquad \phi(p, \mathcal{J}) = (p + p^g_\infty)^{\frac{\mathcal{J}}{A_{g,l}}} (p + p^l_\infty)^{\frac{1}{A_{g,l}}} \tag{114}$$

$$\psi(p, \mathcal{I}, \mathcal{J}) = \sqrt{\frac{\mathcal{J}}{C_m A_{g,l}}} \int^p \frac{\sqrt{R(p, \mathcal{J})}}{(p + p^g_\infty)^{n_1/2}(p + p^l_\infty)^{n_2/2}} dp \tag{115}$$

Proof As developed previously, the differential equations (110) have been used to determine the k-Riemann invariants for the k-characteristic genuinely non-linear. It is found that the k-Riemann invariants are constants across the k-rarefaction waves. Therefore, we can write

$$\begin{cases} \mathcal{J} = \mathcal{J}(\mathcal{U}_l) = \mathcal{J}(\mathcal{U}_r) = \dfrac{\alpha(p + p^g_\infty)}{(1 - \alpha)(p + p^l_\infty)} \\ \mathcal{I} = \mathcal{I}(\mathcal{U}_l) = \mathcal{I}(\mathcal{U}_r) = \dfrac{T}{\phi(p, \mathcal{J})} \\ \mathcal{K}_1(\mathcal{U}_l) = \mathcal{K}_1(\mathcal{U}_r) = u + \psi(p, \mathcal{I}, \mathcal{J}), \qquad for \quad 1 - rarefaction. \end{cases} \tag{116}$$

□

$$\begin{cases} \mathcal{J} = \mathcal{J}(\mathcal{U}_l) = \mathcal{J}(\mathcal{U}_r) = \dfrac{\alpha(p + p^g_\infty)}{(1 - \alpha)(p + p^l_\infty)} \\ \mathcal{I} = \mathcal{I}(\mathcal{U}_l) = \mathcal{I}(\mathcal{U}_r) = \dfrac{T}{\phi(p, \mathcal{J})} \\ \mathcal{K}_4(\mathcal{U}_l) = \mathcal{K}_4(\mathcal{U}_r) = u - \psi(p, \mathcal{I}, \mathcal{J}), \qquad for \quad 4 - rarefaction. \end{cases} \tag{117}$$

Hence the relationship between the states $\mathcal{U}_l = (\alpha_l, p_l, u_l, T_l)^T$ and $\mathcal{U}_r = (\alpha_r, p_r, u_r, T_r)^T$ can be written as

$$\begin{cases} \alpha_r = \varphi(p_r, \mathcal{J}) \\ T_r = T_l \dfrac{\phi(p_r, \mathcal{J})}{\phi(p_l, \mathcal{J})} \\ u_r = u_l - \Big(\psi(p_r, \mathcal{I}, \mathcal{J}) - \psi(p_l, \mathcal{I}, \mathcal{J})\Big), \qquad for \quad 1-rarefaction. \end{cases} \tag{118}$$

And

$$\begin{cases} \alpha_r = \varphi(p_r, \mathcal{J}) \\ T_r = T_l \dfrac{\phi(p_r, \mathcal{J})}{\phi(p_l, \mathcal{J})} \\ u_r = u_l + \Big(\psi(p_r, \mathcal{I}, \mathcal{J}) - \psi(p_l, \mathcal{I}, \mathcal{J})\Big), \qquad for \quad 4-rarefaction. \end{cases} \tag{119}$$

Here the functions $\mathcal{I}, \mathcal{J}, \varphi, \phi$, and ψ are defined by the relations (113)–(115).

To study the admissibility condition of (110): $\lambda_k(\mathcal{U}_l) \le \xi \le \lambda_k(\mathcal{U}_r)$, we denote by $\mathcal{R}_k(\mathcal{U}_l)$ the set of states $\mathcal{U} = (\alpha, p, u, T)^T$ that can be connected from $\mathcal{U}_l$ via a k-rarefaction wave.

Next we use Eqs. (111) and (112) in order to parametrize the vector $\mathcal{U}$ under the form:

$$\mathcal{U}(p) = \begin{pmatrix} \alpha(p) \\ p \\ u(p) \\ T(p) \end{pmatrix} = \begin{pmatrix} \varphi(p, \mathcal{J}) \\ p \\ u_l \pm \Big(\psi(p, \mathcal{I}, \mathcal{J}) - \psi(p_l, \mathcal{I}, \mathcal{J})\Big) \\ T_l \dfrac{\phi(p, \mathcal{J})}{\phi(p_l, \mathcal{J})} \end{pmatrix} \tag{120}$$

As mentioned before in (110), the rarefaction wave $\mathcal{U}_k(\xi)$ is determined by solving the non-linear equation:

$$\lambda_k(\mathcal{U}_k(\xi)) = \xi, \tag{121}$$

ξ is a given real.

By applying the parametrization (120), the resolution of Eq. (121) is reduced to find a parameter p solution of the equation:

$$\lambda_k(\mathcal{U}_k(p)) = \xi \tag{122}$$

To solve this non-linear equation, we have to determine the monotonicity of the following function:

$$p \longrightarrow \lambda_k(\mathcal{U}_k(p))$$

By differentiating this function, we obtain the expression:

$$\frac{d}{dp}\lambda_k(\mathcal{U}_k(p) = \nabla\lambda_k(\mathcal{U}_k(p)).r_k(\mathcal{U}_k(p)) \tag{123}$$

We deduce from (64) that

$$\frac{d}{dp}\lambda_1(\mathcal{U}_k(p) < 0 \quad and \quad \frac{d}{dp}\lambda_4(\mathcal{U}_k(p) > 0$$

This demonstrates that the 1-rarefaction wave is monotonically decreasing and the 4-rarefaction wave is monotonically increasing.

Finally, the admissibility condition $\lambda_k(\mathcal{U}_l) \leq \lambda_k(\mathcal{U}_r)$ is equivalent to

$$\begin{cases} p_r \leq p_l \quad , \quad for \quad 1 - rarefaction. \\ p_l \leq p_r \quad , \quad for \quad 4 - rarefaction \end{cases} \tag{124}$$

9 Rankine–Hugoniot Relation

Let us consider two states $\mathcal{U}_l = (\alpha_l, p_l, u_l, T_l)^T$ and $\mathcal{U}_r = (\alpha_r, p_r, u_r, T_r)^T$ connected by a discontinuity. Then, we have the following Rankine–Hugoniot conditions:

$$[\rho_g \alpha u] = \sigma[\rho_g \alpha] \tag{125}$$

$$[\rho_l(1-\alpha)u] = \sigma[\rho_l(1-\alpha)] \tag{126}$$

$$[\rho u^2 + p] = \sigma[\rho u] \tag{127}$$

$$[Eu + pu] = \sigma[E] \tag{128}$$

where $E = \bar{E} + \dfrac{\rho u^2}{2}$ and σ is the speed of the discontinuity.

Lemma 9.1

Two states $\mathcal{U}_l = (\alpha_l, p_l, u_l, T_l)^T$ and $\mathcal{U}_r = (\alpha_r, p_r, u_r, T_r)^T$ are connected by the Rankine–Hugoniot relations when they satisfy

$$\begin{cases} \alpha_r = \varphi(p_r, \mathcal{J}) \\ \dfrac{1}{\rho_r} = \dfrac{\bar{E}_l + p_m}{\rho_l(\bar{E}_r + p_m)} \\ T_r = \dfrac{\bar{\rho}_m(\alpha_r, p_r)}{\rho_r} \\ u_r = u_l \pm \sqrt{-\Big(p_r - p_l\Big)\Big(\dfrac{1}{\rho_r} - \dfrac{1}{\rho_l}\Big)} \end{cases} \tag{129}$$

Here $\mathcal{J} = \mathcal{J}(\mathcal{U}_l) = \mathcal{J}(\mathcal{U}_r)$ and $p_m = (p_l + p_r)/2$.

Proof By introducing the functions:

$$m_g = \rho_g \alpha(u - \sigma), \quad m_l = \rho_l(1 - \alpha)(u - \sigma), \quad m = \rho(u - \sigma) \tag{130}$$

The Rankine–Hugoniot conditions can be formulated as

$$[m_g] = [m_l] = [m] = 0 \tag{131}$$

$$[u] = m[\frac{1}{\rho}] \tag{132}$$

$$m[u] + [p] = 0 \tag{133}$$

$$m[\frac{E}{\rho}] + [pu] = m[\frac{\bar{E}}{\rho}] + \frac{m}{2}[u^2] + [pu] = 0 \tag{134}$$

To solve these algebraic relations, we distinguish two cases: □

a. Case $m = 0$

It follows from (132) and (133) that

$$u_l = u_r \tag{135}$$

$$p_l = p_r \tag{136}$$

b. Case $m \neq 0$

From Eq. (130), we deduce that

$$[\frac{m_g}{m_l}] = 0 \quad (137)$$

This leads to the following relation:

$$\mathcal{J}(\mathcal{U}) = \mathcal{J}(\mathcal{U}_l) = \mathcal{J}(\mathcal{U}_r) = \frac{\alpha(p + p_\infty^g)}{(1-\alpha)(p + p_\infty^l)} = \frac{\alpha E_l}{(1-\alpha)E_g} \quad (138)$$

Note that $\mathcal{J}(\mathcal{U})$ is identical to the expression defined in (86). Then, we have

$$\alpha_r = \varphi(p_r, \mathcal{J}) \quad (139)$$

where the function $\varphi(p, \mathcal{J})$ is given by (114).

From Eqs. (132) and (133), we deduce an expression for the velocities given by

$$u_r = u_l + m\Big(\frac{1}{\rho_r} - \frac{1}{\rho_l}\Big) \quad (140)$$

where

$$m = \pm\sqrt{-[p]\Big/[\frac{1}{\rho}]} \quad (141)$$

Using Eq. (134), we can also write

$$\frac{m}{2}[u^2] + [pu] = \frac{u_l + u_r}{2} m[u] + p_r[u] + u_l[p] = -\frac{u_l + u_r}{2}[p] + p_r[u] + u_l[p] \quad (142)$$

Then, we get

$$\frac{m}{2}[u^2] + [pu] = -\frac{1}{2}[p][u] + p_r[u] = \frac{p_l + p_r}{2}[u] = m\frac{p_l + p_r}{2}[\frac{1}{\rho}] \quad (143)$$

which yields to

$$[\frac{\bar{E}}{\rho}] + \frac{p_l + p_r}{2}[\frac{1}{\rho}] = 0 \quad (144)$$

This demonstrates that

$$[\frac{1}{\rho}] = -\frac{1}{\rho_l}\frac{1}{\bar{E}_r + p_m}[\bar{E}] = -\frac{1}{\rho_r}\frac{1}{\bar{E}_l + p_m}[\bar{E}] \quad (145)$$

where $p_m = \dfrac{p_l + p_r}{2}$.

Thus, the Rankine–Hugoniot relations between the states $\mathcal{U}_l = (\alpha_l, p_l, u_l, T_l)^T$ and $\mathcal{U}_r = (\alpha_r, p_r, u_r, T_r)^T$ can be written as follows:

$$\begin{cases} \alpha_r = \varphi(p_r, \mathcal{J}) \\ \dfrac{1}{\rho_r} = \dfrac{\bar{E}_l + p_m}{\rho_l(\bar{E}_r + p_m)} \\ T_r = \dfrac{\bar{\rho}_m(\alpha_r, p_r)}{\rho_r} \\ u_r = u_l \pm \sqrt{\dfrac{1}{\rho_l(\bar{E}_r + p_m)}\Big(p_r - p_l\Big)\Big(\bar{E}_r - \bar{E}_l\Big)} \end{cases} \tag{146}$$

Here $\mathcal{J} = \mathcal{J}(\mathcal{U}_l) = \mathcal{J}(\mathcal{U}_r)$ and ρ_k denotes the density of the mixture.

10 Shock Waves

A discontinuity between a left state $\mathcal{U}_l = (\alpha_l, p_l, u_l, T_l)^T$ and a right state $\mathcal{U}_r = (\alpha_r, p_r, u_r, T_r)^T$ is a k-shock wave when they satisfy the Rankine–Hugoniot relations and the admissibility criterion: $\lambda_k(\mathcal{U}_r) \le \sigma \le \lambda_k(\mathcal{U}_l)$, σ denotes the speed shock.

Lemma 10.1

Two states $\mathcal{U}_l = (\alpha_l, p_l, u_l, T_l)^T$ and $\mathcal{U}_r = (\alpha_r, p_r, u_r, T_r)^T$ are joined by a k-shock wave if the following relations are satisfied:

1-Shock Wave

$$\begin{cases} \alpha_r = \varphi(p_r, \mathcal{J}) \\ \dfrac{1}{\rho_r} = \dfrac{\bar{E}_l + p_m}{\rho_l(\bar{E}_r + p_m)} \\ T_r = \dfrac{\bar{\rho}_m(\alpha_r, p_r)}{\rho_r} \\ u_r = u_l - \sqrt{-\Big(p_r - p_l\Big)\Big(\dfrac{1}{\rho_r} - \dfrac{1}{\rho_l}\Big)} \\ p_l \le p_r \end{cases} \tag{147}$$

4-Shock Wave

$$\begin{cases} \alpha_r = \varphi(p_r, \mathcal{J}) \\ \dfrac{1}{\rho_r} = \dfrac{\bar{E}_l + p_m}{\rho_l(\bar{E}_r + p_m)} \\ T_r = \dfrac{\bar{\rho}_m(\alpha_r, p_r)}{\rho_r} \\ u_r = u_l - \sqrt{-\Big(p_r - p_l\Big)\Big(\dfrac{1}{\rho_r} - \dfrac{1}{\rho_l}\Big)} \\ p_r \le p_l \end{cases} \tag{148}$$

where $\mathcal{J} = \mathcal{J}(\mathcal{U}_l) = \mathcal{J}(\mathcal{U}_r)$.

Proof According to the above definition, an 1-shock-wave must satisfy the following conditions:

$$\begin{cases} \alpha_r = \varphi(p_r, \mathcal{J}) \\ \dfrac{1}{\rho_r} = \dfrac{\bar{E}_l + p_m}{\rho_l(\bar{E}_r + p_m)} \\ T_r = \dfrac{\bar{\rho}_m(\alpha_r, p_r)}{\rho_r} \\ u_r = u_l \pm \sqrt{-\Big(p_r - p_l\Big)\Big(\dfrac{1}{\rho_r} - \dfrac{1}{\rho_l}\Big)} \\ \lambda_1(\mathcal{U}_r) \le \sigma \le \lambda_1(\mathcal{U}_l) \end{cases} \tag{149}$$

where

$$\lambda_1(\mathcal{U}) = u - c, \qquad c = \frac{1}{\sqrt{\rho K}}$$

By substituting these expressions into the admissibility condition: $\lambda_1(\mathcal{U}_r) \le \sigma \le \lambda_1(\mathcal{U}_l)$, we obtain the following relations:

$$\begin{cases} m = \rho_r(u_r - \sigma) \le \rho_r c_r \\ \rho_l c_l \le m = \rho_l(u_l - \sigma) \\ m = \rho_l(u_l - \sigma) \ge 0 \end{cases} \tag{150}$$

As $m \geq 0$, the expression of (141) becomes

$$m = \sqrt{-[p] / [\frac{1}{\rho}]} \tag{151}$$

Substituting the expression of m into (150) yields to

$$\rho_l c_l \leq \sqrt{-[p] / [\frac{1}{\rho}]} \leq \rho_r c_r \tag{152}$$

The equivalent form of Eq. (152) is given by

$$\frac{1}{\rho_r^2 c_r^2} \leq -[\frac{1}{\rho}] / [p] \leq \frac{1}{\rho_l^2 c_l^2} \tag{153}$$

Using the definition of the sound speed (24), the above inequalities can be rewritten as

$$\frac{K_r}{\rho_r} \leq -[\frac{1}{\rho}] / [p] \leq \frac{K_l}{\rho_l} \tag{154}$$

Substituting the relations (145) in Eq. (154) yields to

$$K_r(\bar{E}_l + p_m) \leq \frac{[\bar{E}]}{[p]} \leq K_l(\bar{E}_r + p_m) \tag{155}$$

Now, we can use the definitions (25) and (138) to express the term K as

$$K = \frac{a_\alpha - a_p \alpha(1-\alpha)(E_g - E_l)}{\bar{E} + p} = \frac{1}{\bar{E} + p} \frac{d\bar{E}}{dp} \tag{156}$$

Finally, the admissibility condition can be formulated under the following form: □

$$\frac{\bar{E}_l + p_m}{\bar{E}_r + p_r} \frac{d\bar{E}}{dp}(\mathcal{U}_r) \leq \frac{[\bar{E}]}{[p]} \leq \frac{\bar{E}_r + p_m}{\bar{E}_l + p_l} \frac{d\bar{E}}{dp}(\mathcal{U}_l) \tag{157}$$

To demonstrate the inequalities (157), we must use the following lemma:

Lemma 10.2

Let us consider the function Ψ defined by

$$\Psi : p \longrightarrow \bar{E} = \alpha f_g(p) + (1-\alpha) f_l(p) \tag{158}$$

where $\alpha(p) = \varphi(p, \mathcal{J})$.

Then Ψ is monotonic increasing and concave. Furthermore, it satisfies the following proprieties:

- $\lambda_1(\mathcal{U}_r) \leq \sigma \leq \lambda_1(\mathcal{U}_l)$ holds if and only if $p_l \leq p_r$
- $\lambda_4(\mathcal{U}_r) \leq \sigma \leq \lambda_4(\mathcal{U}_l)$ holds is and only if $p_r \leq p_l$

Proof Using the definitions (9) and (114), the differentiation of the function Ψ gives

$$\frac{d\Psi}{dp} = \frac{d\bar{E}}{dp} = a_\alpha - a_p\alpha(1-\alpha)(E_g - E_l) \tag{159}$$

We deduce from (44) that

$$\frac{d\Psi}{dp} = \frac{\mathcal{J}}{E_l + \mathcal{J} E_g} \frac{R(p, \mathcal{J})}{Q(p, \mathcal{J})} = \frac{\mathcal{J} R(p, \mathcal{J})}{\left(\frac{1}{E_g} + \frac{\mathcal{J}}{E_l}\right)^2} \tag{160}$$

By developing this expression, we obtain

$$\frac{d\Psi}{dp} = A + \frac{B}{\left(\frac{1}{E_g} + \frac{\mathcal{J}}{E_l}\right)^2} \tag{161}$$

where

$$\begin{cases} A = \dfrac{1}{(1+\mathcal{J})}\left(\dfrac{\mathcal{J}}{\gamma_g - 1} + \dfrac{1}{\gamma_l - 1}\right) \\[2ex] B = \dfrac{A_{g,l}\mathcal{J}}{1+\mathcal{J}}\left(\dfrac{1}{E_l} - \dfrac{1}{E_g}\right)^2 \end{cases} \tag{162}$$

This demonstrates that

$$\frac{d\Psi}{dp} > 0 \tag{163}$$

Using Eq. (161), the second derivative of Ψ can be derived as follows:

$$\frac{d^2\Psi}{dp^2} = -\frac{2B(1+\mathcal{J})}{\left(\frac{1}{E_g} + \frac{\mathcal{J}}{E_l}\right)^3} \tag{164}$$

Then, we deduce that

$$\frac{d^2\Psi}{dp^2} \leq 0 \tag{165}$$

Finally, Eqs. (163) and (165) confirm that the function Ψ is monotonic increasing and concave.

In order to apply Lemma 10.2, the admissibility condition (157) needs to be written in function of Ψ. This leads to

$$\Psi'(p_r)\frac{\Psi(p_l)+p_m}{\Psi(p_r)+p_r} \leq \frac{\Psi(p_r)-\Psi(p_l)}{p_r-p_l} \leq \Psi'(p_l)\frac{\Psi(p_r)+p_m}{\Psi(p_l)+p_l} \tag{166}$$

a. Case $p_l \leq p_r$

As the function Ψ is monotonic increasing, we have $\Psi(p_l) \leq \Psi(p_r)$ Then

$$\Psi(p_l)+p_l \leq \Psi(p_l)+p_m \leq \Psi(p_r)+p_m \leq \Psi(p_r)+p_r \tag{167}$$

This yields to

$$\frac{\Psi(p_l)+p_m}{\Psi(p_r)+p_r} \leq 1 \leq \frac{\Psi(p_r)+p_m}{\Psi(p_l)+p_l} \tag{168}$$

The monotonic decreasing of the function Ψ' allows to write $0 \leq \Psi'(p_r) \leq \Psi'(p_l)$.
We deduce that

$$\Psi'(p_r)\frac{\Psi(p_l)+p_m}{\Psi(p_r)+p_r} \leq \Psi'(p_r) \leq \frac{\Psi(p_r)-\Psi(p_l)}{p_r-p_l} \leq \Psi'(p_l) \leq \Psi'(p_l)\frac{\Psi(p_r)+p_m}{\Psi(p_l)+p_l} \tag{169}$$

This demonstrates the admissibility condition of (157).

b. Case $p_r \leq p_l$

The monotonic increasing of the function Ψ implies $\Psi(p_r) \leq \Psi(p_l)$ Then,

$$\Psi(p_r)+p_r \leq \Psi(p_r)+p_m \leq \Psi(p_l)+p_m \leq \Psi(p_l)+p_l \tag{170}$$

It follows that

$$1 \leq \frac{\Psi(p_l)+p_m}{\Psi(p_r)+p_r} \quad and \quad \frac{\Psi(p_r)+p_m}{\Psi(p_l)+p_l} \leq 1 \tag{171}$$

Consequently,

$$\Psi'(p_r) \leq \Psi'(p_r)\frac{\Psi(p_l)+p_m}{\Psi(p_r)+p_r} \quad and \quad \frac{\Psi(p_r)+p_m}{\Psi(p_l)+p_l} \leq 1 \tag{172}$$

Suppose that the admissibility relations (166) hold under the condition (172). Then, we have

$$\Psi'(p_r) \leq \Psi'(p_r)\frac{\Psi(p_l)+p_m}{\Psi(p_r)+p_r} \leq \frac{\Psi(p_r)-\Psi(p_l)}{p_r-p_l} \leq \Psi'(p_l)\frac{\Psi(p_r)+p_m}{\Psi(p_l)+p_l} \leq \Psi'(p_l) \tag{173}$$

It follows from the decreasing property of the function Ψ' : $p_l \leq p_r$. This contradicts the original assumption that $p_r \leq p_l$.

We therefore conclude that

$\lambda_1(\mathcal{U}_r) \leq \sigma \leq \lambda_1(\mathcal{U}_l)$ holds if and only if $p_l \leq p_r$

Finally, the formulation of the 1-shock-wave is described as

$$\begin{cases} \alpha_r = \varphi(p_r, \mathcal{J}) \\ \dfrac{1}{\rho_r} = \dfrac{\bar{E}_l + p_m}{\rho_l(\bar{E}_r + p_m)} \\ T_r = \dfrac{\bar{\rho}_m(\alpha_r, p_r)}{\rho_r} \\ u_r = u_l - \sqrt{-\left(p_r - p_l\right)\left(\dfrac{1}{\rho_r} - \dfrac{1}{\rho_l}\right)} \\ p_l \leq p_r \end{cases} \tag{174}$$

Similarly, the 4-shock wave can be defined as

$$\begin{cases} \alpha_r = \varphi(p_r, \mathcal{J}) \\ \dfrac{1}{\rho_r} = \dfrac{\bar{E}_l + p_m}{\rho_l(\bar{E}_r + p_m)} \\ T_r = \dfrac{\bar{\rho}_m(\alpha_r, p_r)}{\rho_r} \\ u_r = u_l \pm \sqrt{-\left(p_r - p_l\right)\left(\dfrac{1}{\rho_r} - \dfrac{1}{\rho_l}\right)} \\ \lambda_4(\mathcal{U}_r) \leq \sigma \leq \lambda_4(\mathcal{U}_l) \end{cases} \tag{175}$$

Following the calculation performed above, we demonstrate that

$$\begin{cases} m = -\sqrt{-[p]\Big/[\dfrac{1}{\rho}]} \leq 0 \\ \rho_r c_r \leq \sqrt{-[p]\Big/[\dfrac{1}{\rho}]} \leq \rho_l c_l \end{cases} \tag{176}$$

This yields to:

$$p_r \leq p_l \quad and \quad [\frac{1}{\rho}] \geq 0 \tag{177}$$

$$\begin{cases} \alpha_r = \varphi(p_r, \mathcal{J}) \\ \dfrac{1}{\rho_r} = \dfrac{\bar{E}_l + p_m}{\rho_l(\bar{E}_r + p_m)} \\ T_r = \dfrac{\bar{\rho}_m(\alpha_r, p_r)}{\rho_r} \\ u_r = u_l - \sqrt{-\Big(p_r - p_l\Big)\Big(\dfrac{1}{\rho_r} - \dfrac{1}{\rho_l}\Big)} \\ p_r \leq p_l \end{cases} \tag{178}$$

where $\mathcal{J} = \mathcal{J}(\mathcal{U}_l) = \mathcal{J}(\mathcal{U}_r)$.

11 Wave Solution

A k-wave between two states $\mathcal{U}_l = (\alpha_l, p_l, u_l, T_l)^T$ and $\mathcal{U}_r = (\alpha_r, p_r, u_r, T_r)^T$ is a jointure that could be

- A k-rarefaction or a k-shock if λ_k is genuinely non-linear.
- A contact discontinuity if λ_k linearly degenerate.

In order to obtain the formulation of the k-wave, we use the functions θ and ω defined by

$$\theta(p, p_0, \mathcal{J}) = \begin{cases} \dfrac{\bar{\rho}_m(\alpha, p)}{\bar{\rho}_m(\alpha_0, p_0)} \dfrac{\bar{E}(\alpha_0, p_0) + (p + p_0)/2}{\bar{E}(\alpha, p) + (p + p_0)/2} & p_0 \leq p (shock) \\ \dfrac{\phi(p, \mathcal{J})}{\phi(p_0, \mathcal{J})} & p \leq p_0 (rarefaction) \end{cases} \tag{179}$$

$$\omega(p, p_0, \mathcal{I}, \mathcal{J}) = \begin{cases} \sqrt{-\left(p - p_0\right)\left(\frac{1}{\rho} - \frac{1}{\rho_0}\right)} & p_0 \le p(shock) \\ \int_{p_0}^{p} \frac{1}{\rho c}\, dp = \psi(p, \mathcal{I}, \mathcal{J}) - \psi(p_0, \mathcal{I}, \mathcal{J}) & p \le p_0(rarefaction) \end{cases} \tag{180}$$

where $\bar{E}$, ρ_m, ϕ, and ψ are, respectively, the functions defined by (11), (17), (114), and (115).

Note that the temperature T and the density of the mixture ρ are computed in function of the pressure according to the relations (111)–(112) for the k-rarefaction, and those defined in (147)–(148) for the k-shock.

Now, the expression of the k-wave that connects a left state $\mathcal{U}_l = (\alpha_l, p_l, u_l, T_l)^T$ to a right state $\mathcal{U}_r = (\alpha_r, p_r, u_r, T_r)^T$ is given as follows:

1-Wave

$$\begin{cases} \alpha_r = \varphi(p_r, \mathcal{J}_l) \\ T_r = T_l \theta(p_r, p_l, \mathcal{J}_l) \\ u_r = u_l - \omega(p_r, p_l, \mathcal{I}_l, \mathcal{J}_l) \end{cases} \tag{181}$$

where $\mathcal{I}_l = \mathcal{I}(\mathcal{U}_l)$, $\mathcal{J}_l = \mathcal{J}(\mathcal{U}_l)$

2-Wave

$$\begin{cases} p_r = p_l \\ u_r = u_l \end{cases} \tag{182}$$

3-Wave

$$\begin{cases} \alpha_l = \varphi(p_l, \mathcal{J}_r) \\ T_l = T_r \theta(p_l, p_r, \mathcal{J}_r) \\ u_l = u_r + \omega(p_l, p_r, \mathcal{I}_r, \mathcal{J}_r) \end{cases} \tag{183}$$

where $\mathcal{I}_r = \mathcal{I}(\mathcal{U}_r), \quad \mathcal{J}_r = \mathcal{J}(\mathcal{U}_r)$.

12 Riemann Problem

In the theory of hyperbolic partial differential equations, the Riemann problem is a system of conservation laws whose initial state is defined by a simple discontinuity composed of two states.

$$\begin{cases} \dfrac{\partial \mathcal{W}}{\partial t} + \dfrac{\partial F(\mathcal{W})}{\partial x} = 0 \\ \mathcal{W}(x, 0) = \begin{cases} \mathcal{W}_l & x < x_0 \\ \mathcal{W}_r & x > x_0 \end{cases} \end{cases} \tag{184}$$

where $\mathcal{W}_l$, $\mathcal{W}_r$ are two given states, $\mathcal{W}$ and $F(\mathcal{W})$ denote the conservative variables defined by (14) and (15) (Fig. 1). According to the general Lax theory, the solution of the Riemann problem (184) gives rise to the existence of two intermediate states $\mathcal{W}_1$, $\mathcal{W}_2$ satisfying the following conditions:

- 1-wave connecting the state $\mathcal{W}_l$ to the state $\mathcal{W}_1$.
- 2-wave connecting the state $\mathcal{W}_1$ to the state $\mathcal{W}_2$.
- 3-wave connecting the state $\mathcal{W}_2$ to the state $\mathcal{W}_r$.

Using the primitive variables $\mathcal{U}_l$, $\mathcal{U}_1$, $\mathcal{U}_2$ and $\mathcal{U}_r$, the previous conditions established for the solution of the Riemann problem can be expressed as follows:

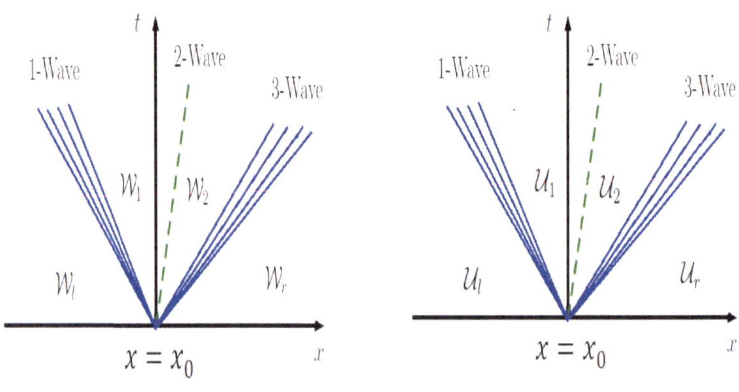

Fig. 1 Riemann solutions with the conservative and primitive variables

$$\begin{cases} \omega(p_*, p_l, \mathcal{I}_l, \mathcal{J}_l) + \omega(p_*, p_r, \mathcal{I}_r, \mathcal{J}_r) + u_r - u_l = 0 \\ u_* = 0.5\Big(u_r - u_l + \omega(p_*, p_r, \mathcal{I}_r, \mathcal{J}_r) - \omega(p_*, p_l, \mathcal{I}_l, \mathcal{J}_l)\Big) \\ \alpha_1 = \varphi(p*, \mathcal{J}_l), \qquad \alpha_2 = \varphi(p*, \mathcal{J}_r) \end{cases} \tag{185}$$

where $\mathcal{I}_k = \mathcal{I}(\mathcal{U}_k)$, $\mathcal{J}_k = \mathcal{J}(\mathcal{U}_k)$ *and* $\mathcal{U}_k = \Big(\alpha_k, p_k, u_k, T_k\Big)^T$, $k = l, r$.

The system (185) indicates that the resolution of the problem of Riemann is reduced to finding the root $p*$ of the non-linear equation:

$$r(p) = \omega(p_*, p_l, \mathcal{I}_l, \mathcal{J}_l) + \omega(p_*, p_r, \mathcal{I}_r, \mathcal{J}_r) + u_r - u_l = 0 \tag{186}$$

Here the function $\omega(p*, p_k, \mathcal{I}_k, \mathcal{J}_k)_{k=l,r}$ is defined in (180).

13 Numerical Experiment

The purpose of this section is to validate the mathematical approach proposed for the resolution of the Riemann problem. To this end, we provide comparative studies by analyzing the following points:

- Validation of the analytical solution of the Riemann problem.
- Treatment of the boundary conditions with the partial Riemann solutions.
- Construction of finite volume methods implementing an exact Riemann solver.

For all the computations, we use a fluid mixture composed of air bubbles dispersed in water. The physical properties of the fluids are summarized in Table 1.

The distance increment is taken to be Δx = L/1000, L being the pipe length. The time increment Δt is evaluated at each iteration through the CFL criterion.

In this section, we begin by giving the definition of the MacCormack scheme, which is one of the most popular numerical schemes known for its second-order accuracy. Then, we give the three comparative studies mentioned before.

Table 1 Parameters of the Stiffened gas EOS for water at T = 355 K

Fluid	γ	$P_\infty(P_a)$	q(J/kg)	$C_p(J/K.kg)$
Liquid	2.35	10^9	$-0.1167\ \ 10^7$	4267
Gas	1.43	0	$0.2030\ \ 10^7$	1487

13.1 MacCormack Scheme

MacCormack's scheme is a two-step Lax–Wendroff-type scheme defined by

Step1: Predictor

$$U_i^* = U_i^n - \frac{\Delta t}{\Delta x}(F_{i+1}^n - F_i^n) \tag{187}$$

Step2: Corrector

$$U_i^{n+1} = \frac{1}{2}(U_i^n + U_i^*) - \frac{\Delta t}{2\Delta x}(F_i^* - F_{i-1}^*) \tag{188}$$

where $F_i^* = f(U_i^*)$ *and* $F_i^n = f(U_i^n)$.

The MacCormack's scheme is known as a conservative scheme in which the numerical flux is given by

$$F_{j+1/2}^n = \frac{1}{2}(f(U_{i+1}^n) + f(U_i^*)) \tag{189}$$

The MacCormack scheme is second-order accurate in space and time. The stability of the scheme is guaranteed by the Courant–Friedrichs–Lewy condition:

$$\frac{\Delta t}{\Delta x} max|\lambda_j| < 1/2 \tag{190}$$

where λ_j is the wave speed.

13.2 Validation of the Exact Riemann Solutions

To validate the robustness of the mathematical analysis developed in the previous sections, we present a comparative study between the numerical results provided by the MacCormack scheme and those obtained analytically for the Riemann problem:

$$\begin{cases} \dfrac{\partial \mathcal{W}}{\partial t} + \dfrac{\partial F(\mathcal{W})}{\partial x} = 0 \\ \mathcal{W}(x,0) = \begin{cases} W_L & x_0 < x \\ W_R & x < x_0 \end{cases} \end{cases} \tag{191}$$

The initial value problem (191) admits an analytical solution composed of constant states separated by k-waves, see Fig. 2.

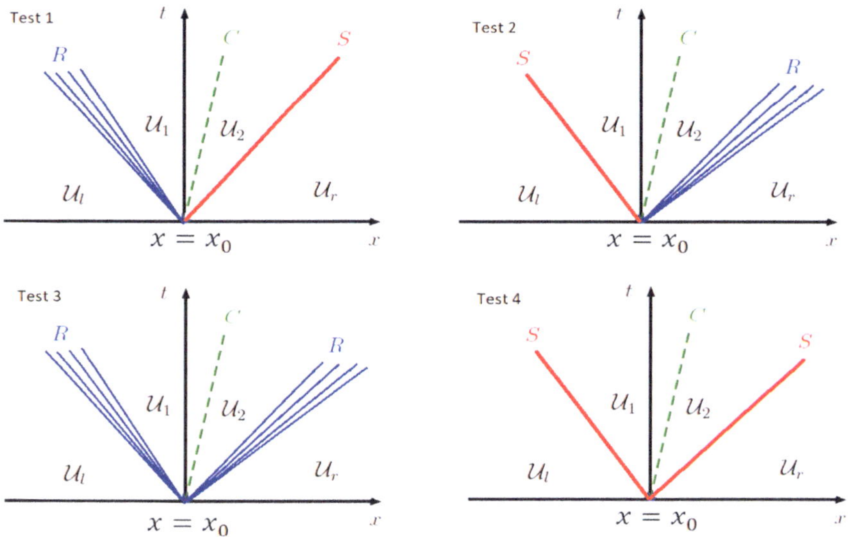

Fig. 2 Riemann solutions: R(Rarefaction), S(Shock), C(Contact discontinuity)

To predict numerically this analytical solution, we consider the four test cases of the Riemann problem, including shock, rarefaction, and contact discontinuity. This allows us to verify the accuracy of the theoretical calculations established for the rarefaction and shock waves, and also to evaluate the entropy property that characterizes the physical solutions. As can be seen in the previous sections, the strong non-linearity of equations makes it difficult to express fluid velocity as a function of pressure. For the rarefaction wave, we have to solve a system of complex differential equations, and for the shock wave, we must solve a system of non-linear algebraic equations.

Figures 3, 4, 5, and 6 present the comparative studies corresponding to the following test cases:

- Test 1: Left rarefaction and right shock waves.
- Test 2: Left shock and right rarefaction waves.
- Test 3: Two rarefaction waves.
- Test 4: Two shock waves.

The initial data of all test cases in terms of primitive variables is given in Table 2.

For the comparative study, we display the results for pressure, velocity, gas volume fraction, and temperature.

The simulation of the Riemann solution gives a good agreement, which has been clearly demonstrated by the different comparisons between the numerical and analytical results.

The numerical results of the MacCormack scheme correctly reproduce the behavior of the physical parameters.

Table 2 Initial conditions

	Test 1	Test 2	Test 3	Test 4
	Rarefaction-Shock	Shock-Rarefaction	Rarefaction-Rarefaction	Shock-Shock
u_*	0.77	0.77	0.77	0.77
p_*	0.5 10^5	0.5 10^5	0.25 10^5	10^5
α_l	5 10^{-5}	5 10^{-5}	5 10^{-5}	5 10^{-5}
p_l	10^5	0.25 10^5	0.5 10^5	0.25 10^5
u_l	$u_* + \omega(p_*, p_l, \mathcal{I}_l, \mathcal{J}_l)$			
T_l	355	355	355	355
α_r	2 10^{-5}	2 10^{-5}	2 10^{-5}	2 10^{-5}
p_r	0.25 10^5	10^5	10^5	0.5 10^5
u_r	$u_* - \omega(p_*, p_r, \mathcal{I}_r, \mathcal{J}_r)$			
T_r	355	355	355	355

The second-order accuracy of this scheme makes the simulation of two-phase flow more efficient. However, it suffers from parasitic oscillations, as shown in Figs. 3, 4, 5, and 6.

13.3 *Treatment of the Boundary Conditions Using the Riemann Problem*

For the hyperbolic system of Equations (1)–(4), the solution of the Riemann problem can be an efficient and robust tool for solving boundary conditions. In this work, we consider the case of boundary conditions consisting of a closure valve at upstream and a reservoir with constant pressure at downstream. The boundary conditions can then be formulated as follows:

- At the upstream side, the velocity of the fluid is taken to be zero.
- At the downstream side, the pressure is kept constant, equal to the reservoir pressure.

Mathematically, the resolution of the above boundary conditions can be achieved by using the analytical solutions of the partial Riemann problem.

13.3.1 Closed Upstream Valve

For a rapid closing valve located at the upstream of the pipe, the mathematical description of this problem can be formulated as follows:

$$\begin{cases} \frac{\partial \mathcal{W}}{\partial t} + \frac{\partial F(\mathcal{W})}{\partial x} = 0, \quad (x,t) \in \Omega, \\ \mathcal{W}(x,0) = \mathcal{W}_0, \\ u(x_0,t) = 0, \end{cases} \tag{192}$$

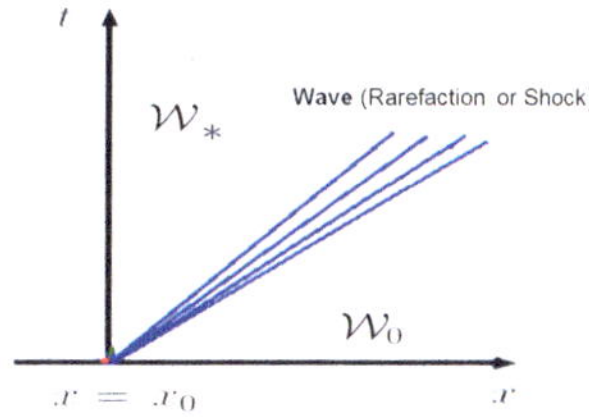

where the initial data is a constant state given by W_0 and the velocity at the boundary $x = x_0$ is imposed to be zero, i.e., $u_0 = 0$. Following [5], the initial boundary value problem (192) is equivalent to the following Riemann problem:

$$\begin{cases} \frac{\partial \mathcal{W}}{\partial t} + \frac{\partial F(\mathcal{W})}{\partial x} = 0, \\ \mathcal{W}(x,0) = \begin{cases} W^-, & x_0 < x, \\ W^+, & x < x_0. \end{cases} \end{cases} \tag{193}$$

Here, the primitive variables of the left and right states W^-, W^+ are defined as functions of the state W_0 according to the following relations:

$$\begin{cases} \alpha^+ = \alpha^- = \alpha_0, \\ p^+ = p^- = p_0, \\ u^+ = -u^- = u_0, \\ T^+ = T^- = T_0. \end{cases} \tag{194}$$

Now, the solution W^* at the boundary condition of (192) is identical to the intermediate state of the Riemann problem (195). This implies that W^* and W_0 are connected by 3-wave (183), which allows to express the primitive variables $(\alpha^*, p^*, u^*, T^*)$ of W^* as follows:

$$\begin{cases} r(p*) = 0, \\ \alpha^* = \varphi(p^*, \mathcal{J}_0), \\ T^* = T_0\theta(p^*, p_0, \mathcal{J}_0) \\ u^* = 0, \end{cases} \tag{195}$$

where $r(p)$ denotes the function defined in (186). The functions $\mathcal{J}$ and $\varphi(p, \mathcal{J})$ are defined by (114).

As expected mathematically, Fig. 7 indicates that the sudden closing valve causes the propagation of a wave of rarefaction if $u_0 > 0$ and of a wave of shock if $u_0 < 0$. The good agreement between the numerical and exact results confirms the robustness of the theoretical approach used for the resolution of the boundary conditions, and proves once again the interest in the use of the exact Riemann solution.

13.3.2 Downstream Reservoir

For a reservoir situated at the downstream, the mathematical problem describing this boundary condition can be formulated as follows.

$$\begin{cases} \dfrac{\partial \mathcal{W}}{\partial t} + \dfrac{\partial F(\mathcal{W})}{\partial x} = 0, \quad (x,t) \in \Omega, \\ \mathcal{W}(x,0) = \mathcal{W}_0, \\ P(x_0,t) = \text{Pressure\_Reservoir}, \end{cases} \quad (196)$$

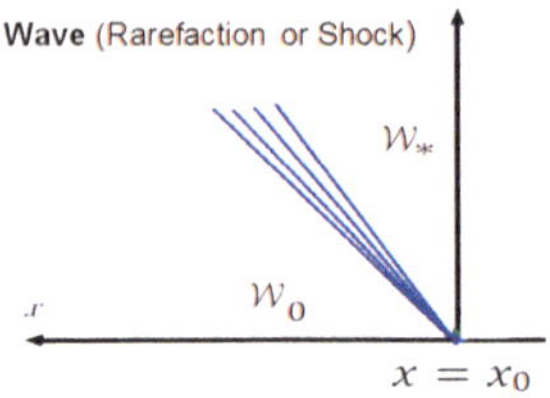

The initial data is a constant state given by W_0 and at the boundary $x = x_0$ the pressure of the fluid is constant, i.e., equal to the reservoir pressure. This case study can be easily solved by applying the notion of partial Riemann problem. Thus, the state W_0 is connected to W_* through a 1-wave function. Following the definition (181), the primitive variables (α^*, p^*, u^*) can be expressed according to:

$$\begin{cases} p_* = \text{Pressure\_Reservoir}, \\ \alpha_* = \varphi(p_*, \mathcal{J}_0), \\ T_* = T_0\theta(p_*, p_0, \mathcal{J}_0) \\ u_* = u_0 - \omega(p_*, p_0, \mathcal{I}_0, \mathcal{J}_0) \end{cases} \quad (197)$$

where the functions $\mathcal{J}$ and $\varphi(p, \mathcal{J})$ are defined by (114). The analytical solution of the partial Riemann problem is similar to the case of closing valve. It takes the formation of a propagation of a Rarefaction wave if $p_* > p_0$, and a shock wave if $p_* < p_0$.

Figure 8 is an interesting case study of the water hammer problem occurring in two-phase gas-liquid flow. The resolution of the boundary conditions was made by implementing a theoretical approach to the partial Riemann problem on both sides: upstream and downstream of the pipe.

Figure 8 also indicates the influence of the mass change on the evolution of the pressure when there is a diffusion of gas caused by a cavitation. Here, we used Henry's law to model the diffusion of gases.

Figure 9 is another example of a downstream closing valve that provides a comparison between our numerical results and those obtained experimentally by Simpson (26). The water hammer problem is one of the future works that can extend the current research.

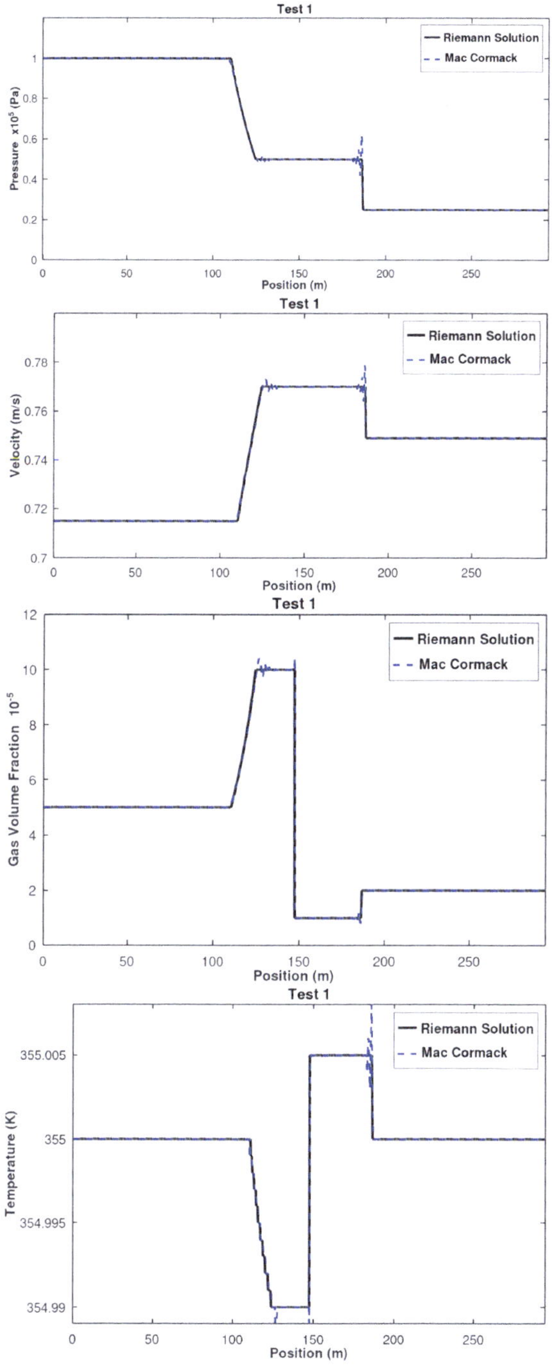

Fig. 3 Comparison between exact and numerical solutions for left rarefaction and right shock waves

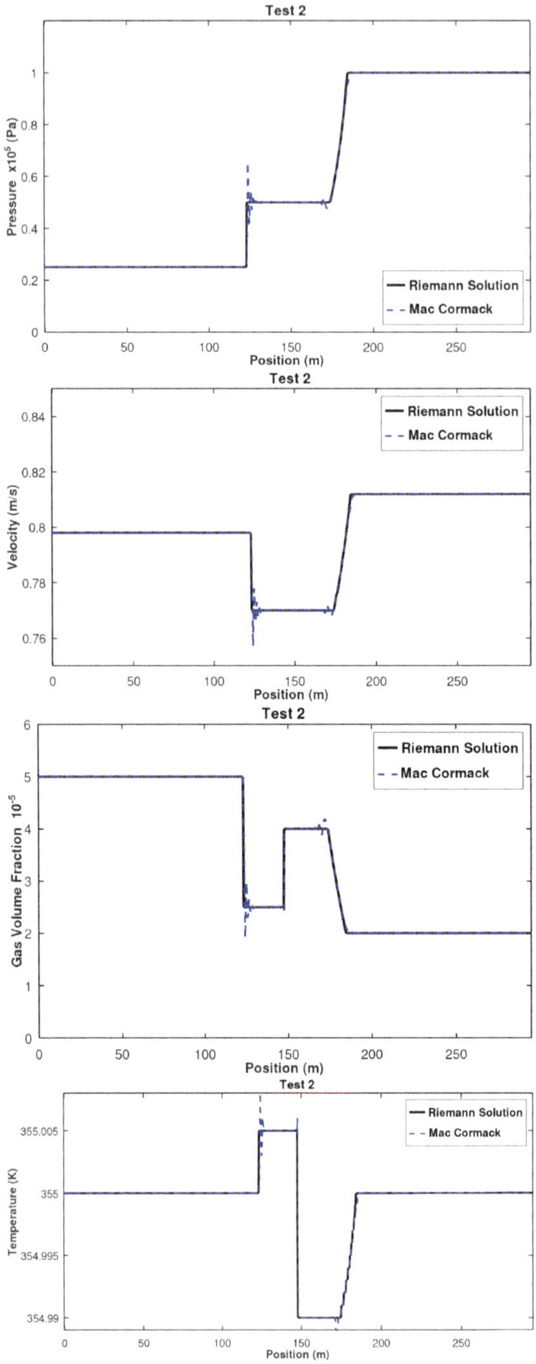

Fig. 4 Comparison between exact and numerical solutions for left shock and right rarefaction waves

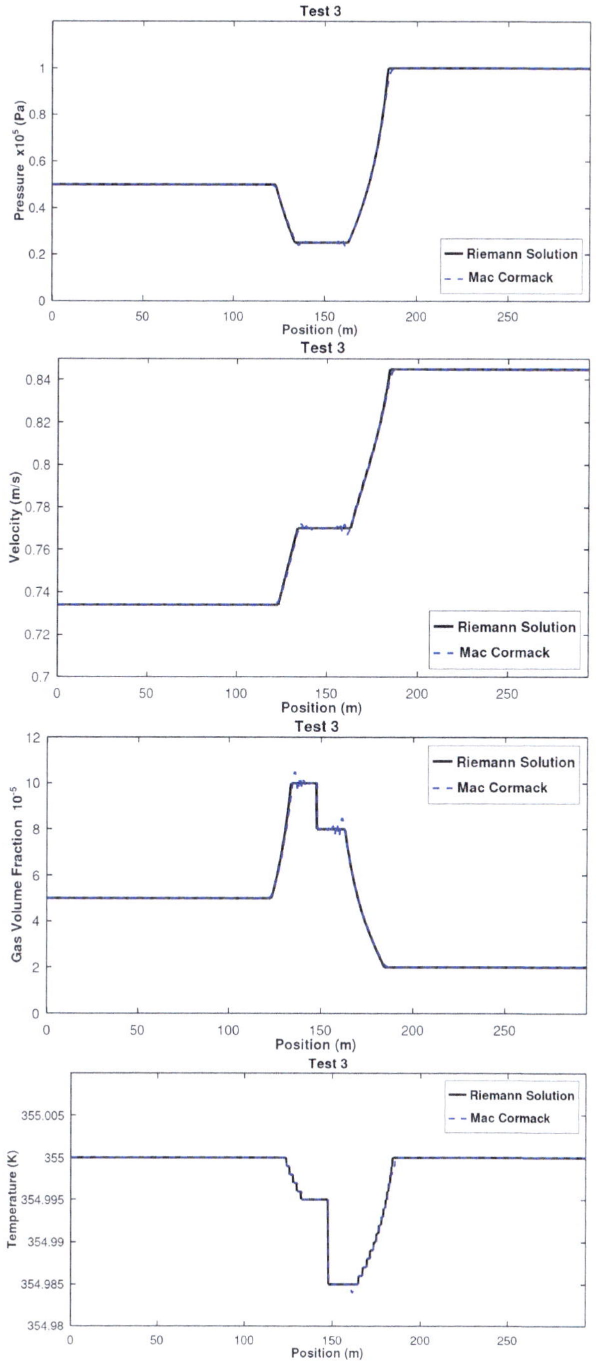

Fig. 5 Comparison between exact and numerical solutions for left and right rarefaction waves

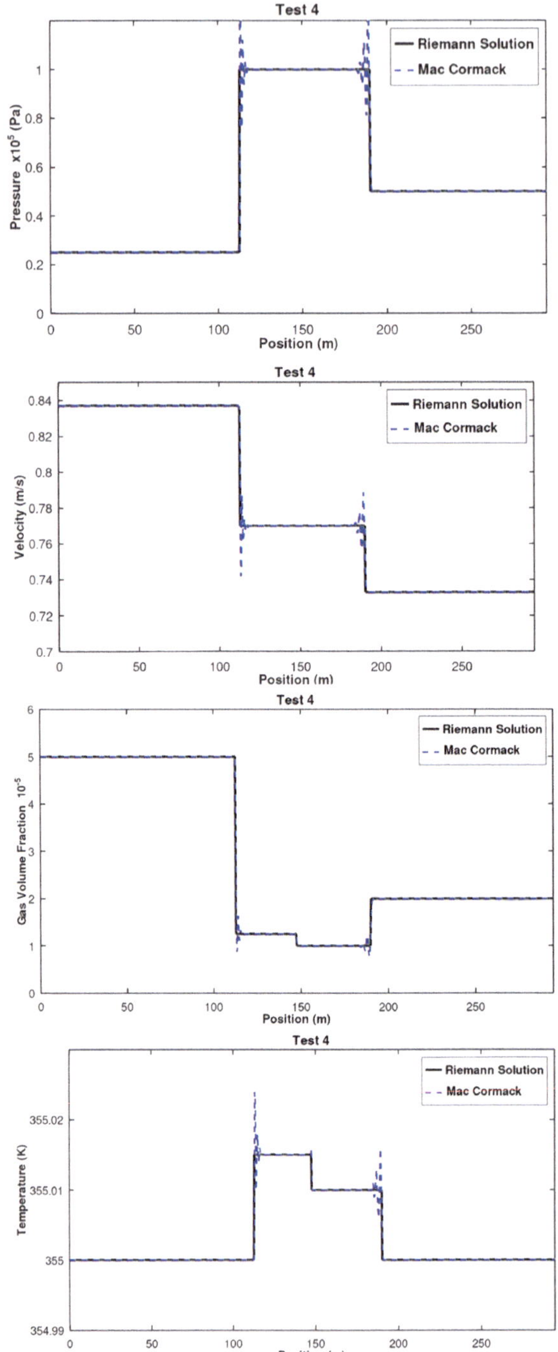

Fig. 6 Comparison between exact and numerical solutions for left and right shock waves

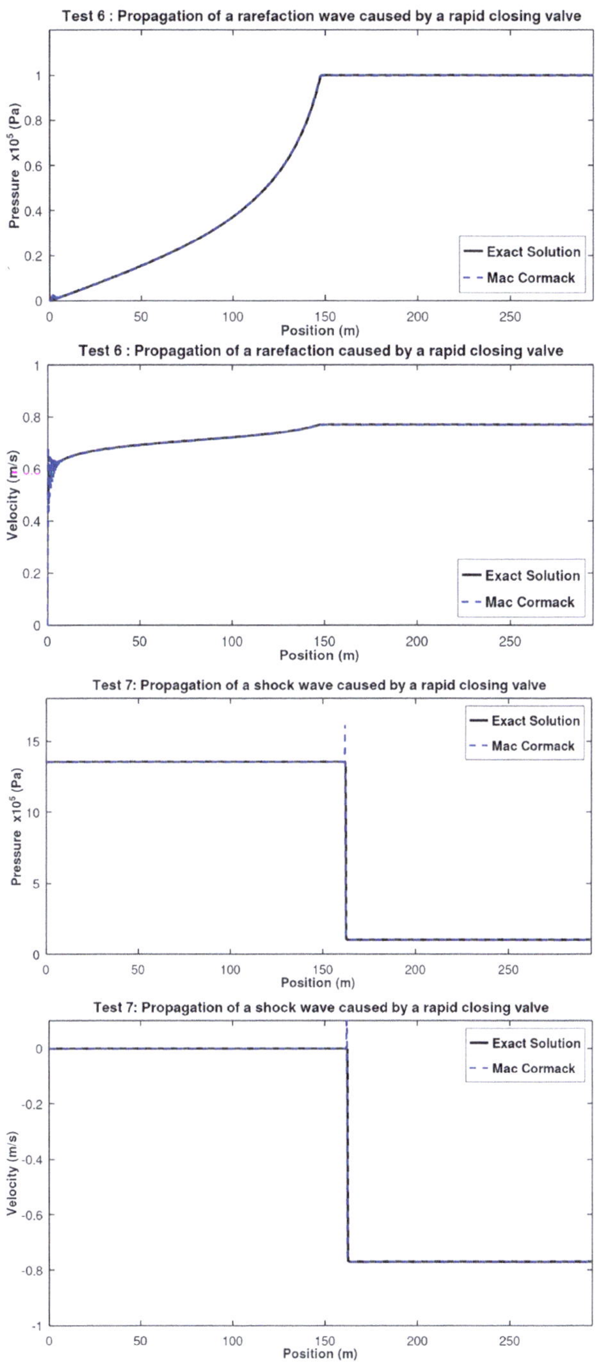

Fig. 7 Treatment of the boundary conditions by using the partial Riemann solution

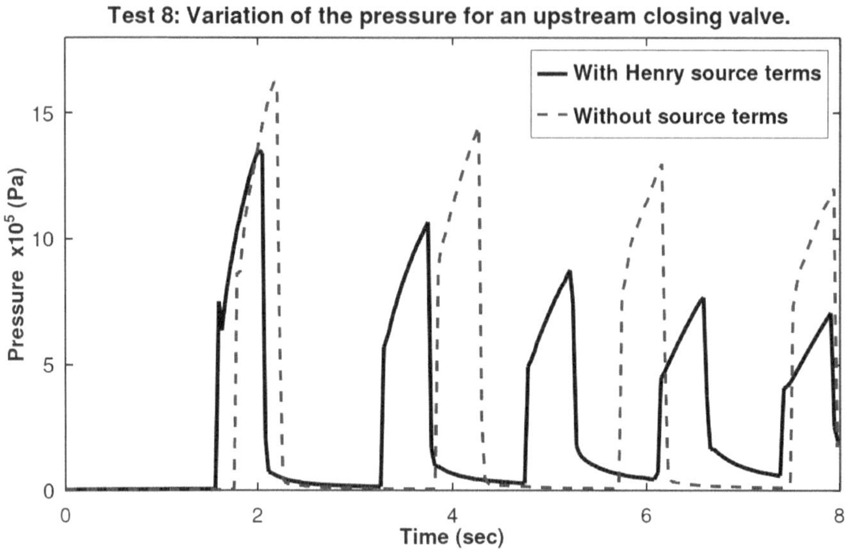

Fig. 8 Variation of the pressure for a water Hammer problem for a pipeline situated between an upstream closed valve and a downstream reservoir with constant pressure

13.4 Construction of Finite Volume Methods Implementing an Exact Riemann Solver

The exact solution of the Riemann problem is the base for the construction of robust and efficient numerical methods such as finite volume methods: Godunov, WAF, TVD-WAF., etc. These schemes give a higher order of precision and can help to eliminate numerical oscillations generated by the presence of shock waves.

In this work, an exact Riemann solver was derived to build Godunov's finite volume method. Figures 10 and 11 shows some comparisons between the exact solutions of the Riemann problem and those obtained by the Godunov method in the cases of tests 1 and 3.

In comparison with MacCormack's scheme, Godunov's results provide the best performance with an absence of numerical oscillations. In our calculation, in the absence of shock, the Godunov methods are convergent for any space increment. However, this is not the case if the Riemann solution contains a shock wave. The construction of finite volume methods and the analysis of the water hammer problem

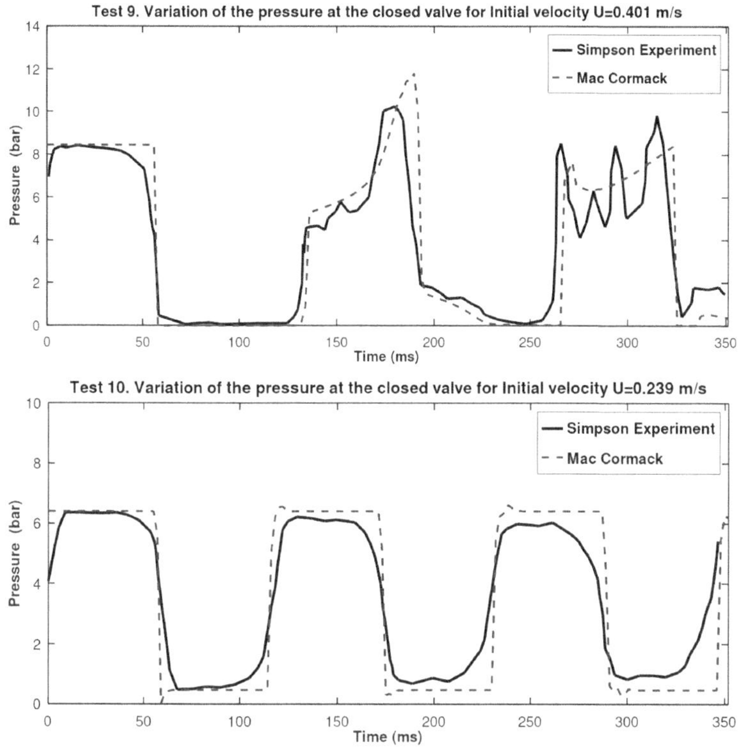

Fig. 9 Comparison between the numerical results and Simpson experiment for a downstream closed valve

are some of the future perspectives of the current work. The results obtained are satisfactory and provide encouraging prospects for future works.

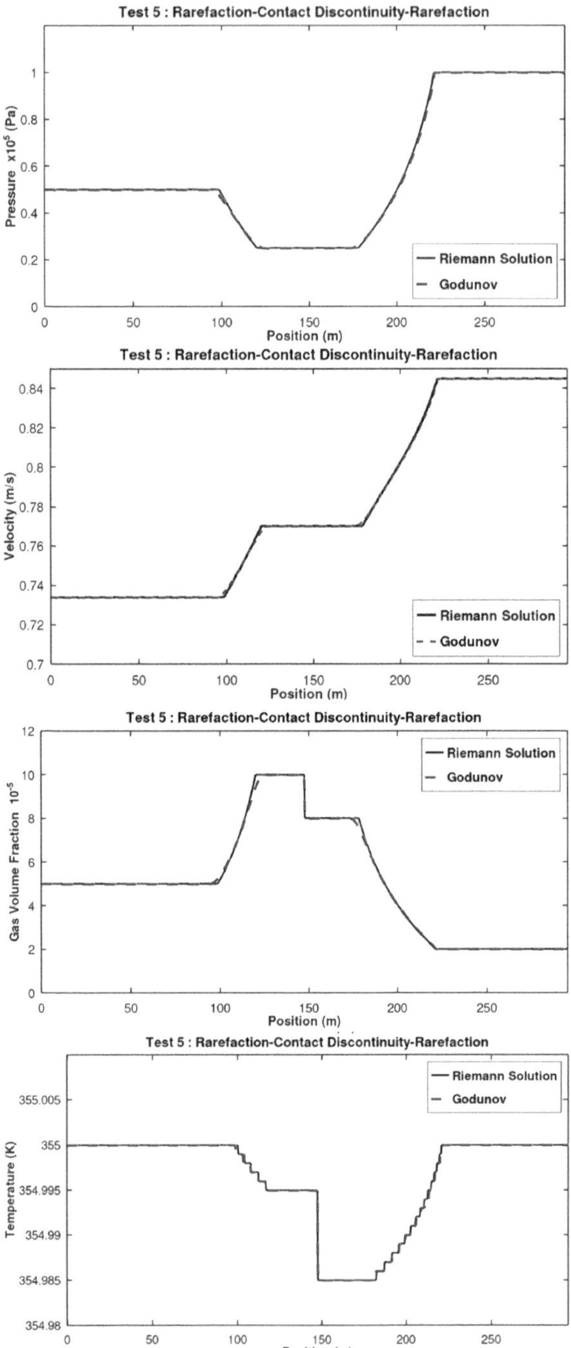

Fig. 10 Comparison between exact and numerical solutions for left and right rarefaction waves

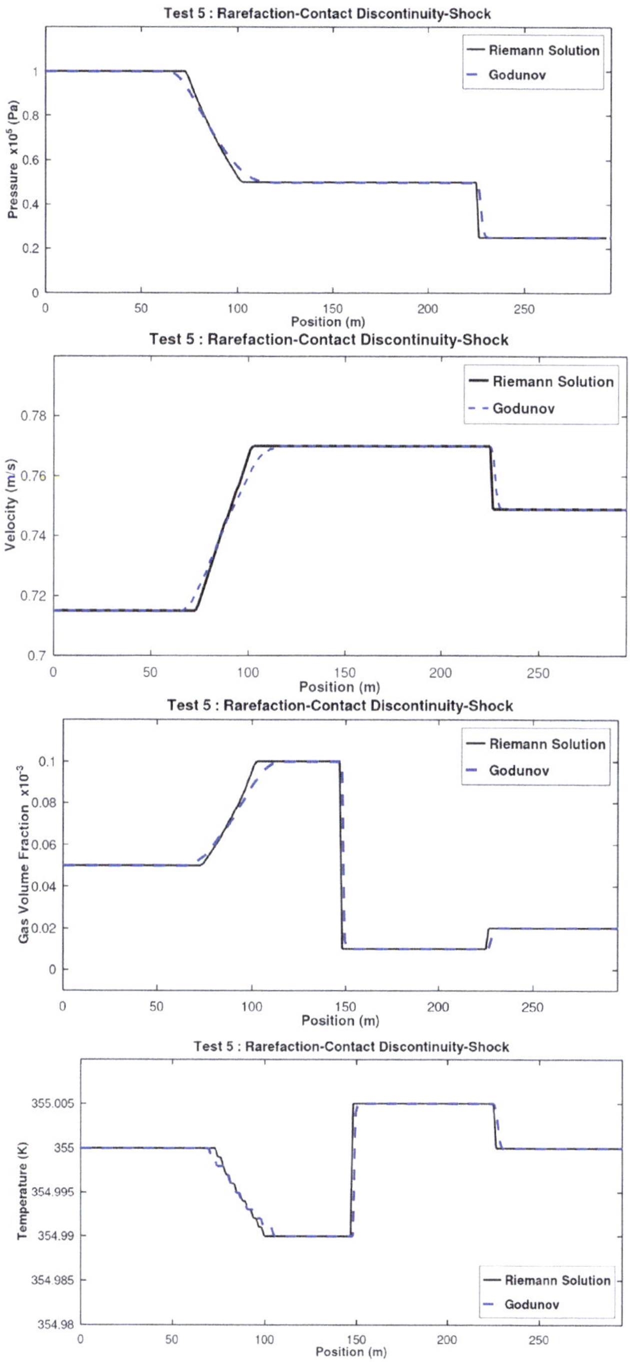

Fig. 11 Comparison between exact and numerical solutions for left rarefaction and right shock waves

14 Conclusion and Future Works

The aim of this research is to provide an analytical solution of the Riemann problem for a four-equation homogeneous equilibrium model. In the literature, this model has been widely used to simulate various physical phenomena such as water hammer, boiling, and condensation. However, few in-depth mathematical studies have been devoted to this model.

In the present work, a rigorous mathematical analysis is presented for the four-equation homogeneous equilibrium model. This includes a novel mathematical approach allowing the exact formulation of shock waves, contact discontinuities, and rarefaction waves. The theoretical correlations obtained for the physical parameters can help to understand the behavior of fluids (gas and liquid) in particular when pressure waves occur inside the flow leading to strong pressure variations.

The validation of the proposed mathematical approach has been demonstrated by the comparisons made between the exact Riemann solutions and the numerical results obtained for different problems: (a) Riemann problem, (b) treatment of boundary conditions with partial Riemann problem, and finally (c) construction of finite volume methods. The exact solution of the Riemann problem can provide useful perspective for further studies, such as

- Construction of robust finite volume methods based on the exact Riemann solver.
- Construction of approximate Riemann solvers.
- Treatment of boundary conditions.
- Simulation of water hammer problems.

References

1. A. Qadi El Idrissi, E. Goncalves, D. Zeidan, Numerical studies of two-phase water hammer flows using Godunov methods. J. Ann. Nuclear Energy **182**, 109578 (2023)
2. D. Zeidan, J. Merker, E. Goncalves Da Silva, L.T. Zhang, *Numerical Fluid Dynamics: Methods and Computations* (Book Springer, 2022)
3. C. Shen, M. Sun, Exact Riemann solutions for the drift-flux equations of two-phase flow under gravity. J. Differ. Equ. **314**, 1–55 (2022)
4. M.D. Thanh, D.X. Vinh, The Riemann problem for the nonisentropic Baer-Nunziato model of two-phase flows. J. Nonlinear Anal.: Real World Appl. **67**, 103623 (2022)
5. A. Qadi El Idrissi, Exact Riemann Solver for a nonlinear hyperbolic system of equations modelling a compressible two-phase flow in pipes. J. Math. Anal. Appl. **505**(1), 125580 (2022)
6. Q. Zhang, The Riemann problem for a drift-flux model of compressible two-phase flow in a variable cross-section duct. Int. J. Non-Linear Mech. **129**, 103644 (2021)
7. F. Daude, A.S. Tijsseling, P. Galon, Numerical investigations of water-hammer with column-separation induced by vaporous cavitation using a one-dimensional Finite-Volume approach. J. Fluids Struct. **83**, 91–118 (2018)
8. M. De Lorenzo, P. Lafon, M. Di Matteo, M. Pelanti, J.-M. Seynhaeve, Y. Bartosiewicz, Homogeneous two-phase flow models and accurate steam-water table look-up method for fast transient simulations. Int. J. Multiph. Flow **95**, 199–219 (2017)

9. I.S. Menshov, Exact and approximate Riemann solvers for compressible two-phase flows. J. Math. Models Comput. Simulat. **9**, 405–422 (2017)
10. M. Dumbser, D.S. Balsara, A new efficient formulation of the HLLEM Riemann solver for general conservative and non-conservative hyperbolic systems. J. Comput. Phys. **304**, 275–319 (2016)
11. D. Zeidan, Assessment of mixture two-phase flow equations for volcanic flows using Godunov-type methods. J. Appl. Math. Comput. **272**(3), 707–719 (2016)
12. S. Kuila, T. Raja Sekhar, D. Zeidan, On the Riemann problem simulation for the drift-flux equations of two-phase flows. Int. J. Comput. Methods **13**(1), 1650009 (2016)
13. H. Lochon, F. Daude, P. Galon, J.-M. Herard, Comparison of two-fluid models on steam-water transients. ESAIM: M2AN **50**, 1631–1657 (2016)
14. S. Kuila, T. Raja Sekhar, D. Zeidan, A robust and accurate Riemann solver for a compressible two-phase flow model. J. Appl. Math. Comput. **265**, 681–695 (2015)
15. F. Crouzet, F. Daude, P. Galon, J.-M. Herard, O. Hurisse, Y. Liu, Validation of a two-fluid model on unsteady liquid-vapor water flows. J. Comput. Fluids **119**, 131–142 (2015)
16. E. Goncalves, Modeling for non isothermal cavitation using 4-equation models. Int. J. Heat Mass Transf. **76**, 247–262 (2014)
17. E. Goncalves, B. Charriere, Modelling for isothermal cavitation with a four-equation model. J. Multiphase Flow **59**, 54–72 (2014)
18. E. Goncalves, Numerical study of expansion tube problems: toward the simulation of cavitation. J. Comput. Fluids **72**, 1–19 (2013)
19. S. Fechter, F. Jaegle, V. Schleper, Exact and approximate Riemann solvers at phase boundaries. J. Comput. Fluids **75**, 112–126 (2013)
20. E. Goncalves, R.F. Patella, Constraints on equation of state for cavitating flows with thermodynamic effects. J. Appl. Math. Comput. **217**(11), 5095–5102 (2011)
21. D. Zeidan, The Riemann problem for a hyperbolic model of two-phase flow in conservative form. Int. J. Comput. Fluid Dyn. **25**(6), 299–318 (2011)
22. A. Ambroso, J.-M. Herard, O. Hurisse, A method to couple HEM and HRM two-phase flow models. J. Comput. Fluids **38**(4), 738–756 (2009)
23. D. Zeidan, E. Romenski, A. Slaouti, E.F. Toro, Numerical study of wave propagation in compressible two-phase flow. Int. J. Numer. Meth. Fluids **54**(4), 393–417 (2007)
24. D. Zeidan, A. Slaouti, E. Romenski, E.F. Toro, Numerical solution for hyperbolic conservative two-phase flow equations. Int. J. Comput. Methods **4**(2), 299–333 (2007)
25. G.-S. Yeom, K.-S. Chang, Numerical simulation of two-fluid two-phase flows by HLL scheme using an approximate Jacobian matrix. J. Numer. Heat Transf. Part B: Fundam. **49**(2), 155–177 (2006)
26. A. Bergant, A.R. Simpson, A.S. Tijsseling, Water hammer with column separation: a historical review. J. Fluids Struct. **22**(2), 135–171 (2006)
27. D.W. Schwendeman, C.W. Wahle, A.K. Kapila, The Riemann problem and a high-resolution Godunov method for a model of compressible two-phase flow. J. Comput. Phys. **212**(2), 490–526 (2006)
28. E.F. Toro, *Riemann Solvers and Numerical Methods for Fluid Dynamics*, 2nd edn. (Springer, New York, 1999)

A Comparison of $k - \varepsilon$—Type Turbulence Models for Prediction of Inclined Dense Jets

Xinyun Wang and Abdolmajid Mohammadian

Abstract As a class of turbulence models in RANS schemes, $k - \varepsilon$ turbulence models are widely used in simulations for engineering applications due to their computational efficiency and adequate accuracy. This study aims to investigate and evaluate the performance of four commonly used $k - \varepsilon$ models—the standard $k - \varepsilon$ model, the realizable $k - \varepsilon$ model, the Renormalization Group (RNG) $k - \varepsilon$ model, and the $q - \zeta$ (q-zeta) model–on the simulation of an inclined dense jet in a stagnant water environment. The important geometrical and dilution characteristics of inclined dense jets, such as terminal rise height, location of centerline peak, horizontal location of return point/impact point, and dilution at these points, were measured and analyzed. The mixing information was also investigated through the cross section and axial direction of the jet. Four turbulence models were analyzed in terms of their ability to simulate the mixing characteristics of inclined dense jets by comparing the results with previous experimental studies. They showed good performance when the discharge angle was smaller than 60°. The $q - \zeta$ model and the realizable $k - \varepsilon$ model show more accurate results than the other two models.

Keywords $k - \varepsilon$ · Turbulence models · Inclined dense jet · OpenFOAM

1 Introduction

In recent decades, due to the constantly increasing population and limited supply of freshwater resources, an increasing number of people and countries have been using desalination plants for daily consumption, significantly relieving water usage stress [26]. The desalination process produces extremely concentrated brine, which can be harmful to the marine environment if improperly discharged. The use of an inclined dense jet at the outfalls is a common method for increasing the mixing efficiency and

X. Wang (✉) · A. Mohammadian
Department of Civil Engineering, University of Ottawa, 161 Louis Pasteur Private, Ottawa, ON K1N 6N5, Canada
e-mail: xwang457@uottawa.ca

D. Zeidan et al. (eds.), *Computational Fluid Dynamics: Novel Numerical and Computational Approaches*, Infosys Science Foundation Series,
https://doi.org/10.1007/978-981-97-8152-2_7

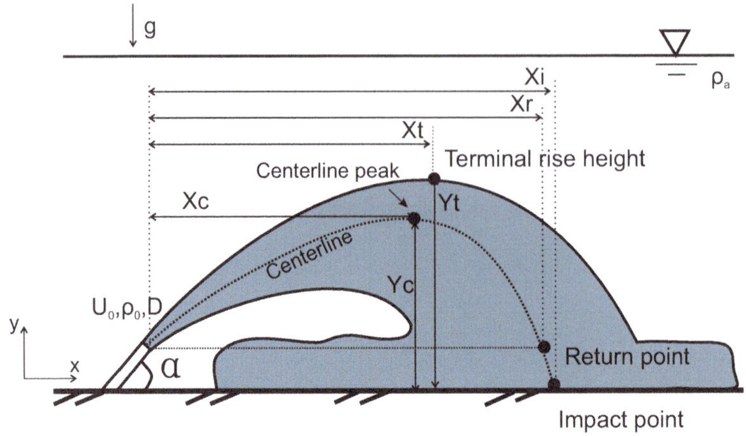

Fig. 1 Schematic of an inclined dense jet discharged into stagnant water

minimizing harmful impacts. Figure 1 shows a typical diagram of a single inclined dense jet discharged into stagnant water. Highly concentrated brine is discharged as a turbulent flow through a submerged pipeline and circular nozzle at a certain inclination angle α. After reaching the highest point, the brine flow descended back to the bottom and continued to move along the receiving floor [29].

The design of the desalination plant outfall system significantly influences brine discharge and mixing. Conversely, various combinations of discharge angle, discharge velocity, temperature, and salinity and environmental factors related to high-concentration brine can also impact or constrain outfall system design. This requires engineer to comprehensively understand the important characteristics related to the discharge and mixing performance, such as the geometrical and dilution information at specific points and mixing development of the flow field. Considering stagnant water as the most challenging receiving condition, because concentrated brine lacks additional forces to aid in the mixing process, extensive research have been conducted over the past few decades to study the characteristics of inclined dense jets under such water conditions [10, 17–28, 37, 38, 40, 46].

Conducting experimental research on inclined dense jets is an effective method to directly yield valuable results. However, it is essential to acknowledge that such studies may encounter certain limitations, particularly concerning the scale of the experimental setup and availability of specialized equipment. For example, advanced techniques, such as particle image velocimetry (PIV) and laser-induced fluorescence (LIF), offer valuable insights into fluid particle movements or flow concentration fields. However, these methods require a high-precision laser and specialized laboratory setup. In addition, the accuracy of the results may be influenced by errors in the instruments and human error. Moreover, adjusting the experiments for different discharge conditions (temperature, concentration level, and discharge angle) can be time-consuming. In contrast, computational fluid dynamics (CFD) has been widely applied and has contributed to engineering applications with the aid of supercomputers in recent decades [22]. It seeks to solve the Navier–Stokes equations using numerical algorithms to capture the essential flow dynamic properties [23]. Current technology and meshing methods can provide accurate flow-field information through direct numerical simulation (DNS). It can solve the smallest turbulent eddies on a sufficiently fine grid for a sufficiently long time and generate instantaneous results that laboratory instruments cannot measure [35] however, DNS is significantly expensive and time-consuming, making it impractical for engineering applications. Therefore, more efficient numerical methods are typically used in practical engineering applications. One approach is to use Reynolds-averaged Navier–Stokes (RANS) equations, which capture the time-averaged behavior of fluid flows by decomposing them into mean and fluctuating component [14]. Several turbulence models have been developed to address the unknown terms in RANS modelling equations to solve them mathematically. Among the different RANS turbulence models, the $k - \varepsilon$ model is one of the most common models for simulating complex turbulent flows. Owing to its reliable results, the $k - \varepsilon$ model is widely used in several industrial fields, such as thermodynamics, maritime navigation, and pollution management [11, 20, 52].

The simulation of inclined dense jets using RANS has been the subject of many studies in the recent decades. Vafeiadou et al. [34] numerically simulated a 60° angled dense jet a Shear Stress transport (SST) turbulence model using the ANSYS software CFX-5. They discovered that the model underestimated the geometrical information compared with the experimental results from Roberts et al. [40]. Gil-deh et al. [17] investigated the performance of five RANs turbulence models in predicting the geometrical and mixing properties of an inclined dense jet with discharge angles of 30° and 45°. The results showed that the LRR and realizable $k - \varepsilon$ models had the most accurate results compared with the experimental results [46]. These studies highlighted that RANS can predict the critical characteristics of inclined dense jets. Moreover, the choice of the turbulence model can affect the RANS results.

In this study, a wide range of discharge conditions, including four different discharge angles (15°, 30°, 45°, 60°) and three Froude numbers (10.6, 14.2, 17.6), will be selected to simulate the behavior of inclined dense jets using four Reynolds-averaged Navier–Stokes (RANS) turbulence models: the standard $k - \varepsilon$ model, realizable $k - \varepsilon$ model, RNG $k - \varepsilon$ model, and the $q - \zeta$ model. The results obtained from these numerical simulations were compared with previous experimental data to evaluate the predictive capabilities of RANS models for inclined dense jets in fully submerged stagnant water. These comparisons provide insights into the performance of these commonly used turbulence models in predicting the mixing properties of inclined dense jets under various conditions.

2 Materials and Methods

2.1 Dimensional Analysis

A typical inclined dense jet is shown in Fig. 1. Highly dense brine (ρ_0) is discharged with an initial velocity U_0 through a round nozzle (D) into the stagnant surrounding water (ρ_a). The inclined jet discharge angle is α.

The inclined dense jet can be characterized by the fluxes of volume Q, momentum M and buoyancy B, and their equations are given by:

$$Q = \frac{\pi}{4} D^2 U_0 \tag{1}$$

$$M = UQ = \frac{\pi}{4} D^2 U^2 \tag{2}$$

$$B = \frac{\pi}{4} D^2 U_0 g_0' \tag{3}$$

where $g_0' = g \frac{(\rho_0 - \rho_a)}{\rho_a}$

The two length scales can be derived from Q, M and B, which are L_Q and L_m. L_Q is the length at which the volume flux of the entrained ambient is equal to the initial volume flux, Q; L_m is the length at which the buoyant flux is equal to the initial momentum flux, M.

These two lengths can also be expressed by the Froude number Fr. Therefore, they can be described as

$$L_m = \frac{M^{3/4}}{B^{1/2}} = \left(\frac{\pi}{4}\right)^{\frac{1}{4}} FrD \tag{4}$$

$$L_Q = \frac{Q}{M^{1/2}} = \left(\frac{\pi}{4}\right)^{\frac{1}{2}} D \tag{5}$$

$$Fr = \frac{U_0}{\sqrt{g_0' D}} \tag{6}$$

According to Pincince [39] and Roberts [40], the effect of the volume flux Q can be negligible, so only l_m length is adopted in the dimensional analysis. Hence, in an inclined dense jet, the characteristic length X can be expressed as:

$$\frac{X}{L_m} = f(L_m, \alpha) \text{ or } \frac{X}{DFr} = f(Fr, \alpha) \tag{7}$$

The functions $f(L_m, \alpha)$ and $f(Fr, \alpha)$ should be obtained from the experiments. One common conclusion drawn from previous research is that for a given discharged angle (α), the (L_m, α) and $f(Fr, \alpha)$ are performed as constants. Such as

$$\frac{X}{L_m} = C_1 or \frac{X}{DFr} = C_1(\text{given } \alpha) \tag{8}$$

For example, the terminal rise height (Y_t) can be expressed in dimensionless form as $\frac{Y_t}{DFr}$, and one constant C_1 can be easily obtained from experiments or numerical simulations.

Similarly, the normalized dilution can also be expressed as:

$$\frac{S}{Fr} = C_2(\text{given } \alpha) \tag{9}$$

where dilution is defined as $S = \frac{C_0 - C_a}{C - C_a}$, where C is the local average concentration. C_0 and C_a are the concentrations in the jet and ambient water, respectively. For example, for a given discharge angle, the normalized dilution at the impact point can be $\frac{S_i}{Fr} = C_2$, similar to Eq. 8, this constant is obtained from experimental studies.

2.2 *Computational Methodology*

2.2.1 Governing Equations

The Navier–Stokes equations are a set of partial differential equations that mathematically describe the motion of a fluid with the basic principles of conservation of mass, momentum, and energy [23]. Most engineering flows are turbulent; however, the description of turbulent flow is always challenging because of its chaotic and random movements with a wide range of time and length scales [13, 42]. Consequently, it is an economical method to describe the motion of all fluid particles in a turbulent flow. Each velocity and all other flow properties were divided into a mean value ($\overline{u}$) and fluctuation components (u'). For example, $u(x, t) = \overline{u}(x) + u'(x, t)$. This is also known as Reynolds decomposition [32]. After applying these divided properties to the Navier–Stokes equation, the Reynolds-averaged Navier (Stokes) can be derived. The governing equations for the Reynolds-averaged Navier–Stokes (RANS) modelling is listed below (Eqs. 10 and 11), including the continuity and momentum equations and no energy equations, because there is no heat transfer in this scope of study.

$$\frac{\partial \rho \overline{u}_i}{\partial i} = 0 \tag{10}$$

$$\frac{\partial \overline{u}_i}{\partial t} + \frac{\partial}{\partial x_j}\left(\overline{u}_i \overline{u}_j\right) = -\frac{1}{\rho}\frac{\partial \overline{P}}{\partial x_i} + v\frac{\partial^2 \overline{u}_i}{\partial x_i x_j} - \frac{\partial \overline{u_i' u_j'}}{\partial x_j} \tag{11}$$

It should be noted that additional unknown terms (Reynolds stress $\overline{u_i'u_j'}$) appear in the momentum equation after the Reynolds decomposition in Eq. 11. This Reynolds stress tensor is diagonally symmetric and contains six unknowns; therefore, it is necessary to construct models for Reynolds stress to close the RANS equations. These models are known as turbulence models.

Applying the Boussinesq hypothesis to the unknown Reynolds stress tensor $\overline{u_i'u_j'}$ [21, 43] the Reynolds stress can be written as:

$$-\overline{u_i'u_j'} = \nu_t \left(\frac{\partial \overline{u}_i}{\partial x_j} + \frac{\partial \overline{u}_j}{\partial x_i} - \frac{2}{3} \frac{\partial \overline{u}_k}{\partial x_k} \delta_{ij} \right) - \frac{2}{3} k \delta_{ij} \tag{12}$$

where k is the turbulence kinetic energy, defined as: $k = \frac{1}{2}\left(\overline{u'u'} + \overline{v'v'} + \overline{w'w'}\right)$ where u', v', w' are the fluctuation for the components of velocity. (δ_{ij} is the Kronecker delta ($\delta_{ij} = 1$ if $i = j$ and $\delta_{ij} = 0$ if $i \neq j$). ν_t is the turbulent eddy viscosity that is strongly related to the local turbulence intensity.

After applying the Boussinesq hypothesis to Eq. 11, the momentum equation in the RANS model can be written as:

$$\frac{\partial u_i}{\partial t} + \frac{\partial}{\partial x_j}\left(u_i u_j\right) = -\frac{1}{\rho}\frac{\partial p}{\partial x_i} + \frac{\partial}{\partial x_j}\left[(\nu_t + \nu)\frac{\partial u_i}{\partial x_j} - \frac{2}{3} k \delta_{ij} \right] \tag{13}$$

There are two unknowns in Eq. 13, ν_t and k. The modeling problems switch from calculating the Reynolds stress ($\overline{u_i'u_j'}$) to the eddy viscosity (ν_t). As a result, the eddy viscosity models are obtained by solving additional partial differential equations, and they are classified according to the number of equations to be solved [23]: zero-equation models, one-equation models, two-equation models. In this study, only two-equation models, which are the most widely used and validated in industrial practice, were adopted and analyzed. These two-equation models can also be classified generally as $k - \varepsilon$ or $k - \omega$ models. In this study, we only considered the $k - \varepsilon$ type models.

In the standard two-equation $k - \varepsilon$ model, eddy viscosity is expressed as follows:

$$\nu_t = \frac{C_\mu k^2}{\varepsilon} \tag{14}$$

where C_μ denotes a dimensionless constant, k denotes the turbulent kinetic energy, and ε denotes the dissipation rate. Both can be solved using transport equations. In this section, some $k - \varepsilon$ type models are used as described below. More information is available in Anderson and Wendt [23].

The general form of the transport equation for k and ε is presented below:

Transport equation for k:

$$\frac{D(\rho k)}{Dt} = P_k - \Phi_k + D_k^\mu + D_k^T \tag{15}$$

Transport equation for ε

$$\frac{D(\rho\varepsilon)}{Dt} = P_\varepsilon - \Phi_\varepsilon + D_\varepsilon^\mu + D_\varepsilon^T \tag{16}$$

In the terms on the right side P is the production term, Φ_k is the destruction term, D_k^μ, D_k^T are the viscous dissipation term and turbulent dissipation terms, respectively. These terms will change in different $k - \varepsilon$ models.

Standard $k - \varepsilon$ Equation

The standard $\mathrm{k} - \varepsilon$ equation is one of the most common forms with the following transport equation for the k and ε [2] (all the models are for incompressible flow).

k equation:

$$\frac{\partial k}{\partial t} + \frac{\partial(ku_i)}{\partial x_i} = -\rho\overline{u_i'u_j'}\frac{\partial u_j}{\partial x_i} + \frac{\partial}{\partial x_j}\left(\left(v + \frac{v_t}{\sigma_k}\right)\frac{\partial k}{\partial x_j}\right) - \varepsilon \tag{17}$$

ε Equation:

$$\frac{\partial \varepsilon}{\partial t} + \frac{\partial(\varepsilon u_i)}{\partial x_i} = \frac{\partial}{\partial x_j}\left(\left(v + \frac{v_t}{\sigma_\varepsilon}\right)\frac{\partial \varepsilon}{\partial x_j}\right) + C_{\varepsilon 1}\frac{\varepsilon}{k}\left(-\rho\overline{u_i'u_j'}\frac{\partial u_j}{\partial x_i}\right) - C_{\varepsilon 2}\frac{\varepsilon^2}{k} \tag{18}$$

The equations have five constants: $C_\mu = 0.09, \sigma_\varepsilon = 1.3, \sigma_k = 1, C_{\varepsilon 1} = 1.44, C_{\varepsilon 2} = 1.92$. The viscosity can be calculated from Eq. 14.

Realizable $k - \varepsilon$

The difference between the standard Realizable $k - \varepsilon$ equations is the equation for ε and the constant for C_μ. In the Realizable $k - \varepsilon$ model, the C_μ is no longer constant. Therefore, the ε equation is given by:

ε Equation:

$$\frac{\partial \varepsilon}{\partial t} + u_i\frac{\partial \varepsilon}{\partial x_i} = \frac{\partial}{\partial x_j}\left(\left(v + \frac{v_t}{\sigma_\varepsilon}\right)\frac{\partial \varepsilon}{\partial x_j}\right) + C_1 S\varepsilon - C_2\frac{\varepsilon^2}{k + \sqrt{v\varepsilon}} \tag{19}$$

where the constant C_1 in Eq. 19 is

$$C_1 = \max\left\{0.43, \frac{\eta}{\eta + 5}\right\}, \eta = \frac{Sk}{\varepsilon}, S = \sqrt{2S_{ij}S_{ij}} \tag{20}$$

where C_2=1.9, σ_ε=1.2.

The constant for C_μ Eq. 14 is written as

$$C_\mu = \frac{1}{A_0 + A_s\frac{u^* k}{\varepsilon}} \tag{21}$$

where $A_0 = 4.04, A_s = \sqrt{6}\cos\phi$

$$\phi = \frac{1}{3}\arccos\left(\sqrt{6}\mathrm{W}\right),\ W = \frac{S_{ij}S_{jk}S_{ki}}{\tilde{S}^3},\ \tilde{S} = \sqrt{S_{ij}S_{ij}},\ S_{ij} = \frac{1}{2}\left(\frac{\partial u_i}{\partial x_j} + \frac{\partial u_j}{\partial x_i}\right)$$

$$u^* = \sqrt{S_{ij}S_{ij} + \tilde{\Omega}_{ij}\tilde{\Omega}_{ij}} \quad (22)$$

where the $\tilde{\Omega}_{ij}$ is the mean rate-of rotation tensor.

RNG $k - \varepsilon$ (Renormalization Group)

Yakhot and Smith [47] proposed the Renormalization Group (RNG). The two transport equations are as follows:

K-Equation:

$$\frac{\partial k}{\partial t} + \frac{\partial (ku_i)}{\partial x_i} = \frac{\partial}{\partial x_j}\left(\left(v + \frac{v_t}{\sigma_k}\right)\frac{\partial k}{\partial x_j}\right) + v_t S^2 - \varepsilon \quad (25)$$

ε-Equation:

$$\frac{\partial \varepsilon}{\partial t} + \frac{\partial (\varepsilon u_i)}{\partial x_i} = \frac{\partial}{\partial x_j}\left(\left(v + \frac{v_t}{\sigma_\varepsilon}\right)\frac{\partial \varepsilon}{\partial x_j}\right) + C_{\varepsilon 1}\frac{\varepsilon}{k}\left(v_t S^2\right) - C^*_{\varepsilon 2}\left(\frac{\varepsilon^2}{k}\right) \quad (26)$$

where $v_t = C_\mu \frac{k^2}{\varepsilon}$, $C^*_{\varepsilon 2} = C_{\varepsilon 2} + \frac{C_\mu \rho \eta^3\left(1 - \frac{\eta}{\eta_0}\right)}{1 + \beta\eta^3}$, $\eta = \frac{k}{\varepsilon}\overline{S}$, $\overline{S} = \sqrt{2\overline{S_{ij}}\overline{S_{ij}}}$, $\overline{S_{ij}} = \frac{1}{2}\left(\frac{\partial \overline{u_i}}{\partial x_j} + \frac{\partial \overline{u_j}}{\partial x_i}\right)$, $S \equiv \sqrt{2S_{ij}S_{ij}}$, $S_{ij} = \frac{1}{2}\left(\frac{\partial u_i}{\partial x_j} + \frac{\partial u_j}{\partial x_i}\right)$

Compared with the standard $k - \varepsilon$ model, there is one extra term in ε equation, which could make the RNG $k - \varepsilon$ model have a wider range of applications and higher accuracy in high-Reynold turbulence flow.

***q* − ζ Turbulence Model**

The $q - \zeta$ turbulence model improves the model's performance near the walls, by adopting damping functions based on the flow type [30]. Strictly speaking, $q - \zeta$ does not belong to the traditional k-epsilon turbulence model, but the q and ζ in the model are expressed as $q = \sqrt{k}$ and $\zeta = \frac{\varepsilon}{2q}$. Eddy viscosity v_t can be defined in terms of $v_t = \frac{C_\mu f_\mu q^3}{2\zeta}$, f_μ is the damping function near the wall, which is defined differently based on the flow type. The $q - \zeta$ model has a similar form to the $k - \varepsilon$ model, thereby considering it a k-epsilon-type model. The $q - \zeta$ equations are as follows:

q-Equation

$$u_i\frac{\partial q}{\partial x_i} = \frac{\partial}{\partial x_j}\left\{\left(v + \frac{v_t}{\sigma_q}\right)\frac{\partial q}{\partial x_j}\right\} + Q - \zeta \quad (27)$$

ζ-Equation

$$u_i\frac{\partial \varsigma}{\partial x_i} = \frac{\partial}{\partial x_j}\left\{\left(v + \frac{v_t}{\sigma_\varsigma}\right)\frac{\partial \varsigma}{\partial x_j}\right\} + \frac{\varsigma}{q}\left(C_{\varsigma 1}f_{\varsigma 1}Q - C_{\varsigma 2}f_{\varsigma 2}\varsigma\right) + \psi' + \xi' \quad (28)$$

$$\psi' = \psi / 2q$$

where ψ is an additional source term that represents an additional production by mean shear to achieve correct limiting behavior very close to a wall. Q is the rate of production of q

$$Q \equiv \frac{P}{2q} = \nu_t \frac{\partial u_i}{\partial x_j}\left(\frac{\partial u_i}{\partial x_j} + \frac{\partial u_j}{\partial x_i}\right) \quad (29)$$

The basic constants are the same as those of the standard $k - \varepsilon$ mode, with additional parameters. $C_\mu = 0.09, C_{\zeta 1} = 1.44, C_{\zeta 2} = 1.92, \sigma_q = 1.0 \text{ and } \sigma_\zeta = 1.3. f_{\zeta 1} = 1.0, f_{\zeta 2} = 1 - 0.3\exp(-Rt^2), R_t = \frac{q*k}{2*\nu*\zeta}$.

The damping function of f_μ is different, the equation is listed as follows:

For anisotropic flow

$$f_\mu = \exp\left[\frac{-2.5 + \frac{Rt}{20}}{\left(1 + \frac{Rt}{130}\right)^3}\right] \quad (30)$$

For isotropic flow

$$f_\mu = \exp\left[\frac{-6.0}{\left(1 + \frac{Rt}{50}\right)^2} * \left(1 + 3.0 * \exp\left(-\frac{Rt}{10}\right)\right)\right] \quad (31)$$

The reader can refer to Gibson and Adil [30] and OpenFOAM user's guide for more model details [6].

2.2.2 Flow Configuration and Computational Domain Setup

OpenFOAM is an open-source computational fluid dynamics (CFD) software known for its exceptional flexibility, various solvers, compatibility with different operating systems, and remarkable postprocessing capabilities [6]. These features make it a particularly effective tool for addressing the complex challenges in fluid dynamics, including heat transfer and chemical reaction simulations [18]. Consequently, it is widely employed in academic and research settings to simulate various fluid dynamics problems. In this study, Reynolds-averaged Navier–Stokes (RANS) equations were discretized using the finite volume method (FVM), and the transport equations from the precious section were solved using an incompressible solver called "*twoLiquidMixingFoam*". This solver was specifically built to predict the mixing of two incompressible flows, and its exceptional performance and accuracy have led to its widespread adoption in numerous studies related to inclined dense jets [1, 19, 24, 48, 49, 50].

Figure 2a shows a schematic of the simulated rectangular tank with dimensions of 2 m × 0.8 m × 0.8 m. The aim is to simulate the situation in a real experimental study. The jet was discharged from one side and positioned at the center of the tank to mitigate the influence of the front and back wall boundaries (Fig. 2a). The mesh for the simulation was generated using *snappyHexMesh* mesh generator, which employs an adaptive mesh method to refine regions of interest. As shown in Fig. 2b, a box-shaped refinement region with dimensions of 1.8 m × 0.4 m × 0.6 m was established inside. This refined mesh was aimed at improving the accuracy of the velocity and concentration predictions in the near-field zone, and additional refinement was applied near the bottom area, as shown in Fig. 2c.

The diameter of the inlet nozzle (D) is 0.02m, and the center of the nozzle tip is fixed at 0.05m(Y_0) with a positive inclination angle. The inlet and outlet boundary types were set as patches, the atmosphere as symmetrical, and the other types as walls. A standard wall function is used. The inlet boundary conditions are $u = U_0 \times \cos(\theta)$, $v = U_0 \times \sin(\theta)$, $w = 0$, $k = 0.006u^2$, and $\varepsilon = 0.06u^3/D$, where U_0 is the velocity of the jet discharge, u, v, and w are the velocities for the x, y, and z-axes, respectively. The flow rate is fixed at $U = 0.65$ m/s. Simulations are carried out for four different discharge angles = 15°, 30°, 45°, 60° and three different Froude numbers (10.6, 14.2, and 17.6). The flow is assumed to be incompressible, and the jet and ambient water densities are calculated using Eq. 32 proposed by Millero and Poisson [12]. The density of the surrounding water was set as 998.21 kg/m^3 with at a temperature

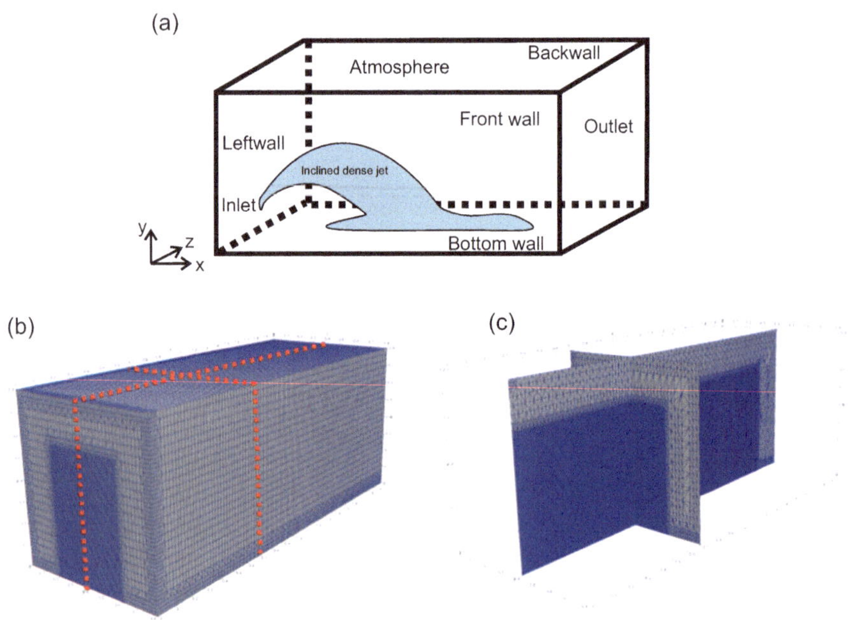

Fig. 2 **a** Schematic figure of computational domain setup; **b** Computational domain with refined mesh; **c** example of mesh on $z-y$ plane and $x-y$ plane

Table 1 Initial condition of the simulation

Velocity (m/s)	Froude number	Density deference (kg/m^3)	Jet density (kg/m^3)
0.65	10.6	19.11	1017.32
0.65	14.2	10.71	1008.92
0.65	17.6	6.81	1005.02

of 20 °C. The initial conditions are listed in Table 1.

$$\rho = \rho_t + AS + BS^{\frac{3}{2}} + CS \quad (32)$$

ρ_t is the density of water, which varies with the temperature as follows:

$$\rho_t = 999.842594 + 6.793952 \times 10^{-2}T - 9.095290 \times 10^{-3}T^2 + 1.001685 \times 10^{-4}$$
$$T^3 - 1.120083 \times 10^{-6}T^4 + 6.536336 \times 10^{-9}T^5$$
$$\mathrm{A} = 8.24493 \times 10^{-1} - 4.0899 \times 10^{-3}T + 7.6438 \times 10^{-5}T^2 - 8.2467 \times 10^{-7}$$
$$T^3 + 5.3875 \times 10^{-9}T^4$$
$$\mathrm{B} = -5.72466 \times 10^{-3} + 1.0227 \times 10^{-4}T - 1.6546 \times 10^{-6}T^2 C = 4.8314 \times 10^{-4}$$

3 Results and Discussion

Forty-eight simulations were carried out with three different Froude numbers and four discharge angles (15°, 30°, 45°, 60°) using Realizable and standard $k-\varepsilon$ turbulence, RNG $-\varepsilon$ and $q-\zeta$ models. Each simulation consistently produced results for over 90 s to ensure that the jet reached the bottom and underwent full stabilization.

3.1 Mesh Quality and Independence Analysis

The mesh was generated using snappyHexMesh, employing a refining box area to precisely capture the flow characteristics of the jet. It is important to note that mesh quality significantly influences the accuracy and reliability of CFD simulations. Although a finer and smaller grid can yield more precise simulation results, it significantly increases the demand for computational resources and simulation time. Hence, selecting an appropriate mesh size within the computational domain selecting an appropriate mesh size within the computational domain. In this study, mesh test runs were conducted using two different meshes: a moderate mesh with two million grids and a fine mesh with four million grids. Two meshes were employed for a specific case involving a 45° discharge angle and a Froude number of 17.6 using standard *k*-epsilon models. Some of the important details of the two generated meshes are listed in Table 2.

The concentration values were obtained along two vertical lines ($x = 5D$ and $x = 10D$) with the jet flow region for both meshes, and the concentration value were plotted against their corresponding vertical locations (y-axis), as illustrated in Fig. 3.

The results indicate that the fine mesh (4 million grinds) did not significantly improve the simulation results but let do a three-fold increase in simulation time. Consequently, for all simulations in this study, a moderate mesh (2 million grids) was adopted to maintain a balance between computational efficiency and accuracy.

Table 2 Mesh quality information

Mesh	Num. of cells	Max aspect ratio	Max skewness	Min volume	Max volume	Mesh non-orthogonality
Moderate mesh	2,131,224	5.39	0.68	1.25e−07	0.00012	Average: 3.9 Max: 56.2
Fine mesh	4,269,931	6.91	0.466	5.10e−09	0.00012	Average: 4.3 Max: 45.7

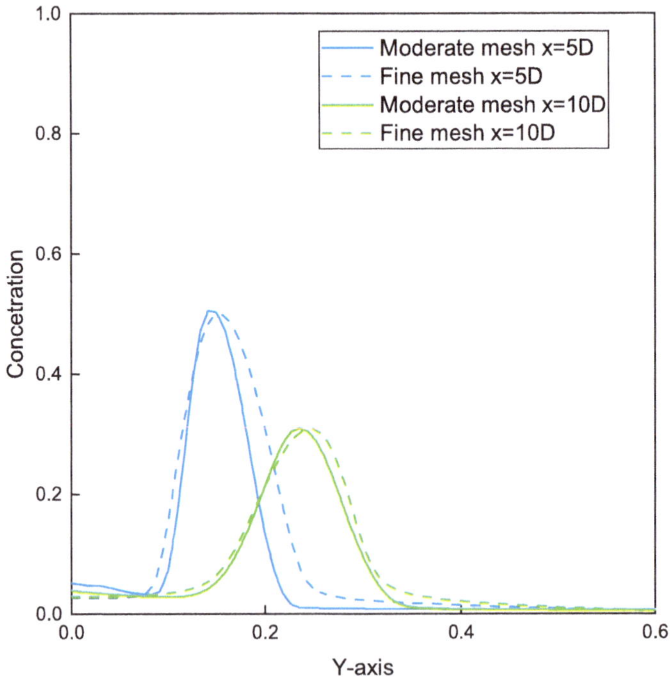

Fig. 3 The concentration comparison between two mesh

3.2 Geometrical and Dilution Information

Understanding the geometrical characteristics of inclined dense jets is critical for designing effective outfalls with minimal environmental impacts. These geometrical characteristics include the terminal rise height (Y_t) and horizontal location of the return (X_r) or impact point (X_i). From the dimensionless analysis presented in Sect. 2, the dimensionless geometrical (X/DFr) and dilution information (S/Fr) of the jet for a given discharge angle (α) remain constant (Eqs. 8 and 9). The relevant coefficients from previous studies, including the results of this study, are summarized in Tables 3, 4, 5 and 6. This section provides a table for a better comparison, and a detailed analysis of the geometrical and dilution information is presented in the following section.

The results demonstrate that the four turbulence models effectively predict the geometrical information of the inclined dense jets. However, when the discharge angle was 60°, an under-prediction of geometrical information was observed. Nevertheless, it is worth noting that the dilution information remained within an acceptable range, highlighting the ability of the turbulence model to capture relevant parameters.

3.2.1 Terminal Rise Height

Similar to vertical negative jets, inclined dense jets ascend until the vertical component of the initial momentum drops to zero, leading to the jet falling back to the bottom due to gravity after reading the maximum height [8]. This highest point is called the terminal rise height, which in our study is defined as Y_t based on the coordinate system setting.

Understanding this height is crucial, as it helps the brine jet from touching the water, as the interaction with the surface can significantly affect the jet's characteristics [46].

Two approaches can be employed to determine the terminal rise height. According to Jirka [25], the visual boundary of turbulent buoyant jets can be defined as the $0.25C_{\max}$ of a concentration centerline value based on the analysis of the experimental intermittency distribution [7, 9]. Another method involves estimating the terminal rise height from the $3\%C_0$ (C_0 is the initial concentration of the jet) cut-off contour of the concentration data [46]. Both methods were tested, and the differences were negligible. For brevity, this study presents the results obtained using the $3\%C_0$ cutoff method. Figure 4. presents the normalized terminal rise height (Y_t/DFr) plotted against the discharge angle (α), and compares with previous experimental studies [3, 4, 27, 28, 46, 29]. The results indicate that a larger discharge angle leads to a higher terminal rise height for the jet, suggesting that larger discharge angles may not be suitable for shallow-water conditions. The agreement between the results and previous experimental studies demonstrated the reliability of the four turbulence models.

Table 3 Comparison of geometrical and mixing coefficients (discharge angle $\alpha = 15°$)

Quantity	Terminal rise height	Horizontal location of centerline peak	Vertical location of centerline peak	Horizontal location of return point	Horizontal location of impact point	Centerline peak dilution	Return point dilution	Impact point dilution
	Y_t/DFr	X_m/DFr	Y_m/DFr	X_r/DFr	X_i/DFr	Sm/Fr	Sr/Fr	Si/Fr
Standard $k - \varepsilon$	0.43	1.11	0.17	1.83	2.42	0.22	0.38	0.46
Realizable $k - \varepsilon$	0.43	1.20	0.18	1.92	2.51	0.21	0.38	0.31
RNG $k - \varepsilon$	0.45	1.05	0.18	1.78	2.33	0.22	0.45	0.52
q-zeta	0.5	1.50	0.17	1.93	2.41	0.21	0.38	0.46
Lai and Lee [28]-Exp	0.44	1.22	0.21	2.41		0.27		0.43
Lai and Lee [28]-VISJET	0.44	1.19	0.21	2.15		0.24		0.41
Kikkert [15] (LIF/LA)	0.52/0.52	1.47/1.22	0.21/0.21	2.53/2.13				

Table 4 Comparison of geometrical and mixing coefficients (discharge angle $\alpha = 30°$)

Quantity	Terminal rise height	Horizontal location of centerline peak	Vertical location of centerline peak	Horizontal location of return point	Horizontal location of impact point	Centerline peak dilution	Return point dilution	Impact point dilution
	Y_t/DFr	X_m/DFr	Y_m/DFr	X_r/DFr	X_i/DFr	Sm/Fr	Sr/Fr	Si/Fr
Standard $k-\varepsilon$	0.76	1.39	0.47	2.36	2.80	0.29	0.57	0.62
Realizable $k-\varepsilon$	0.82	1.47	0.51	2.47	2.82	0.31	0.59	0.63
RNG $k-\varepsilon$	0.80	1.2	0.48	2.28	2.55	0.30	0.66	0.70
q-zeta	0.83	1.61	0.54	2.61	2.84	0.26	0.58	0.66
Lai and Lee [28]-Exp	0.95					0.40		0.82
Lai and Lee [28]-VISJET	0.86					0.31		0.64
Shao and Law [46]	1.05	1.70	0.66	2.88/3.0		0.62/0.66	1.18/1.45	
Kikkert [27] (LIF/LA)	1.19/1.00	1.85/1.75	0.66/0.56	3.44/3.14				
Cipollina et al. [10]	1.08	1.95	0.79	3.03				

Table 5 Comparison of geometrical and mixing coefficients (discharge angle $\alpha = 45°$)

Quantity	Terminal rise height	Horizontal location of centerline peak	Vertical location of centerline peak	Horizontal location of return point	Horizontal location of impact point	Centerline peak dilution	Return point dilution	Impact point dilution
	Y_t/DFr	X_m/DFr	Y_m/DFr	X_r/DFr	X_i/DFr	Sm/Fr	Sr/Fr	Si/Fr
Standard $k-\varepsilon$	1.30	1.32	1.1	2.4	2.65	0.35	0.70	0.75
Realizable $k-\varepsilon$	1.35	1.33	1.05	2.3	2.7	0.34	0.71	0.76
RNG $k-\varepsilon$	1.36	1.25	1.05	2.08	2.26	0.37	0.75	0.79
q-zeta	1.46	1.51	1.11	2.53	2.7	0.4	0.76	0.82
Lai and Lee [28]-Exp	1.58	2.09	1.19	3.34		0.45		1.09
Lai and Lee [28]-VISJET	1.33	1.58	1.00	2.78		0.32		0.77
Shao and Law [46]	1.47	1.69	1.14	2.83		0.46	1.26	
Kikkert [27]	1.60	1.84	1.06	3.26				
Papa-konstantis [37, 38]	1.59		1.17	3.16		0.52		1.55
Cipollina et al. [10]	1.61	1.80	1.17	2.82				

Table 6 Comparison of geometrical and mixing coefficients (discharge angle $\alpha = 60°$)

Quantity	Terminal rise height	Horizontal location of centerline peak	Vertical location of centerline peak	Horizontal location of return point	Horizontal location of impact point	Centerline peak dilution	Return point dilution	Impact point dilution
	Y_t/DFr	X_m/DFr	Y_m/DFr	X_r/DFr	X_i/DFr	Sm/Fr	Sr/Fr	Si/Fr
Standard $k-\varepsilon$	1.53	0.9	1.3	1.62	1.80	0.36	0.76	0.88
Realizable $k-\varepsilon$	1.55	0.82	0.95	1.5	1.62	0.36	0.75	0.81
RNG $k-\varepsilon$	1.46	0.74	1.2	1.42	1.53	0.36	0.84	0.88
q-zeta	1.63	0.93	1.35	1.75	1.88	0.35	0.87	0.91
Lai and Lee [28]-Exp	2.08	1.78	1.64	2.84		0.44		1.07
Lai and Lee [28]-VISJET	1.78	1.34	1.42	2.40		0.29		0.83
Papa-konstantis [37, 38]	2.15		1.68	2.75		0.56		1.68

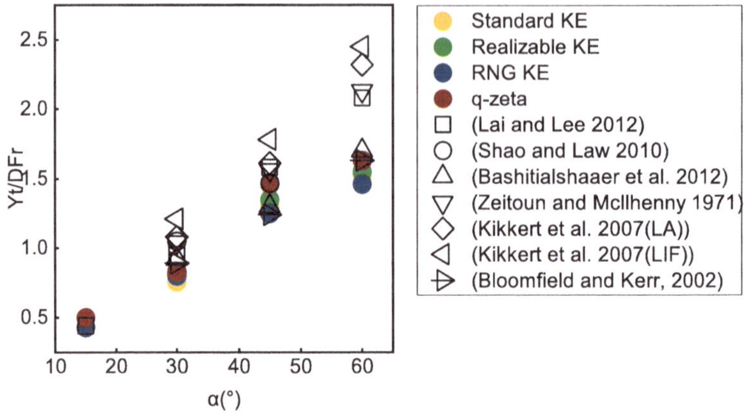

Fig. 4 Normalized terminal rise height plotted against discharge angle

3.2.2 Location of the Return Point and Impact Point

The centerline peak can be determined from the highest point of centerline trajectory, where is the maximum velocity or concentration along the jet flow [17]. The horizontal and vertical locations of the centerline peak are denoted by X_c and Y_c, respectively. The values are normalized and plotted against the discharge angle (α) in Fig. 5. The results were compared with those of previous studies. It can tell that the horizontal location for X_c was under-predicted compared with the experimental study, especially at the 45° and 60°. Underprediction can also be observed for the vertical height Y_c at a 60° angle. This indicates that these four models exhibit relatively poor performance for the larger discharge angles (45° and 60°). For moderate and small discharge angles, $q-\zeta$ model showed the best agreement among the four models.

To investigate the influence of the Froude number, the horizontal location of the centerline peak was normalized with the nozzle diameter and plotted against the Froude number, as shown in Figs. 6 and 7. The comparison was only carried out for the 30° and 45° discharge angles because of the comparable previous results [10, 15, 46]. The results show that the normalized X_c and Y_c exhibit an increasing linear relationship with the Froude number within a certain range. The results from $q-\zeta$ have the best agreement among the four turbulence models.

In the jet's descending path, two points have gained the attention of researchers: the return point and the impact point. Typically, the return point (X_r) is the point where the flow returns to the discharge source height (Y_0) (Fig. 1), whereas the location where the flow impinges on the boundary is called the impact point or impingement point (X_i). One common conclusion that can be agreed upon is that the mixing of the jet is reduced owing to impingement on the bottom [46].

Although these two points are similar from a jet geometrical perspective, the impact point is highly site-specific and depends on the discharge height or the bottom

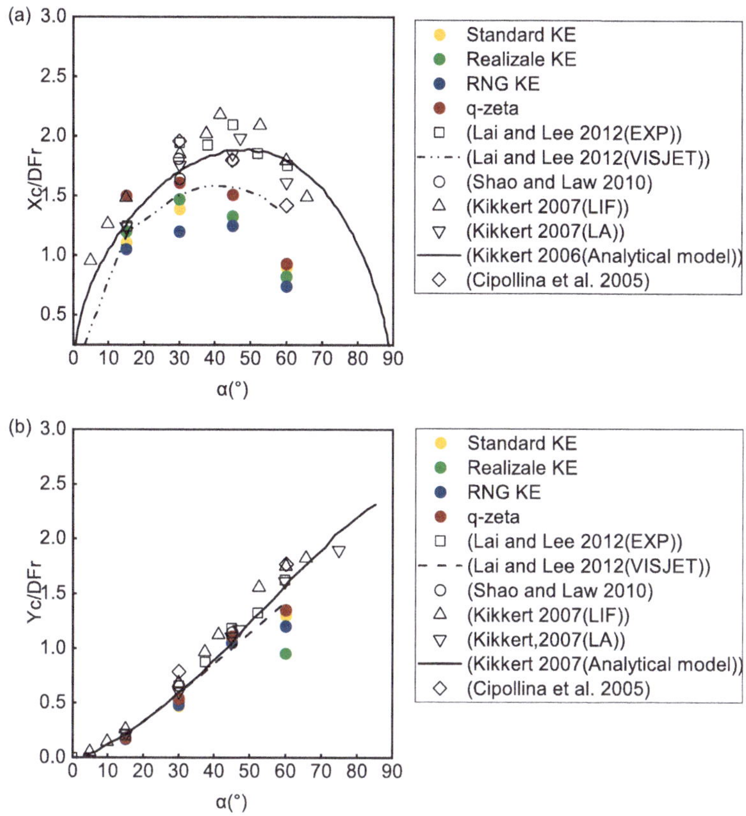

Fig. 5 Normalized centerline peak: **a** horizontal location; **b** vertical location Y_c as a function of discharge angle

condition (sloping or horizontal). More information about the effect of the bottom condition can be found in Oliver et al. [33], Wang and Moham-madian [48]. As a result, some studies only report the return point (X_r) [5, 46], the impact point in Cipollina et al. [10] is referred to as the return point. The return and impact points are interchangeable in Kikkert's study [15]. Shao and Law [46] addressed the importance of distinguishing between return and impact points. Because the return point is independent of the source height and bottom condition, it has consistently been considered a significant parameter rather than the impact point. The concept of return points was also explored by Cipollina et al. [10] and Lai and Lees [28]. Conversely, in the study conducted by Nemlioglu and Roberts [41], the emphasis leaned more toward practical issues associated with impact points, because the impact point can offer insights into interactions with the physical bottom. Therefore, it is necessary to separate and discuss them based on this definition.

In the experimental study, the point of return was determined based on trajectory analysis [46]. However, there are instances where the trajectory cannot be fully

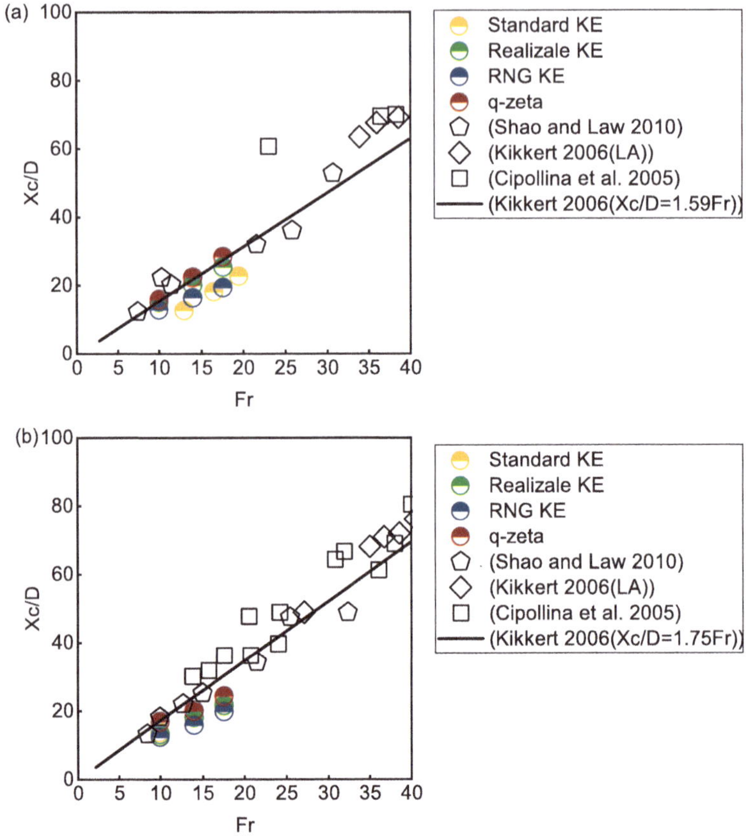

Fig. 6 Normalized horizontal location (X_c) of centerline peak versus Froude number: **a** discharge angle 30°; **b** discharge angle 45

captured as the jet flow returns to the source level due to limitations in the equipment measurements. According to Roberts et al. [40], the impact point can be determined by the variation in the concentration along the lower boundary of the horizontal bottom. Therefore, in this numerical study, the return point was derived from the highest concentration in the horizontal slices at the discharge elevation ($Y_0 = 0.05$ m) (white dashed line in Fig. 8a). The concentration results are plotted along the x-axis in Fig. 8b. The highest value (far from the discharge nozzle) in the concentration profile indicates the location of the return point (Fig. 8b). The impact point was determined using a similar approach involving plotting the concentration against the x-axis location. The point of the highest concentration was the impact point.

In this study, we collected, normalized, and plotted the locations of the return and impact points against the discharge angle (α) for different turbulence models, as shown in Fig. 9. The colored circles represent the impact points, whereas the return points are hollow circles.

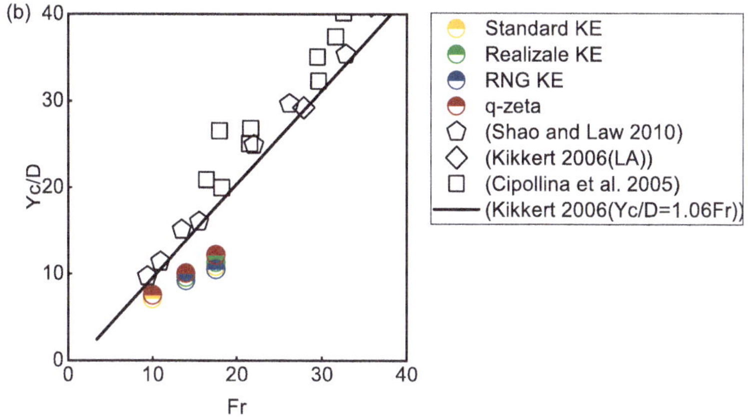

Fig. 7 Normalized vertical location (Y_c) of centerline peak versus Froude number: **a** discharge angle 30°; **b** discharge angle 45°

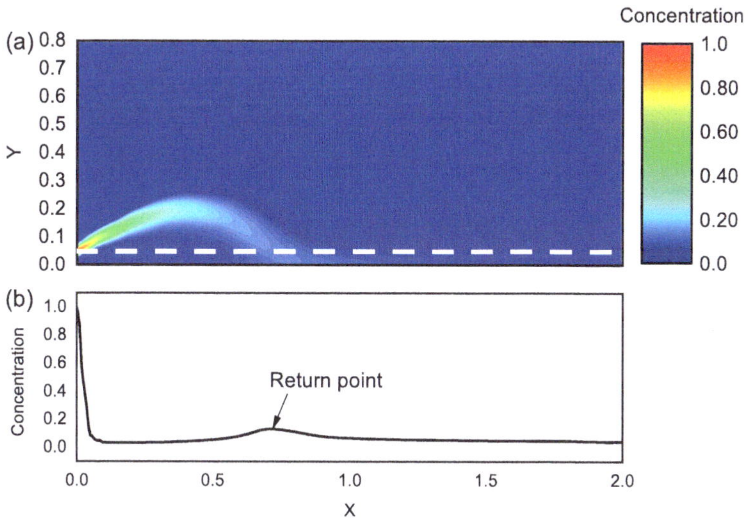

Fig. 8 Schematic diagrams of the concentration: **a** Concentration contours; **b** variation on the horizontal slice shown by a white dashed line at the source elevation. The arrow points out at the return point

Clearly, the variables X_r/DFr and X_i/DFr exhibit a correlation with the discharge angle (α). The jet reaches its maximum when the discharge angle is within the range of $30^\circ \sim 45^\circ$. This observation aligns with the findings of previous experimental studies [10, 27]. It is worth noting that the value of the impact point (solid color) is greater than that of the return point (hollow color) because the source height provides an additional trajectory for movement. This result also indicates that the impact point

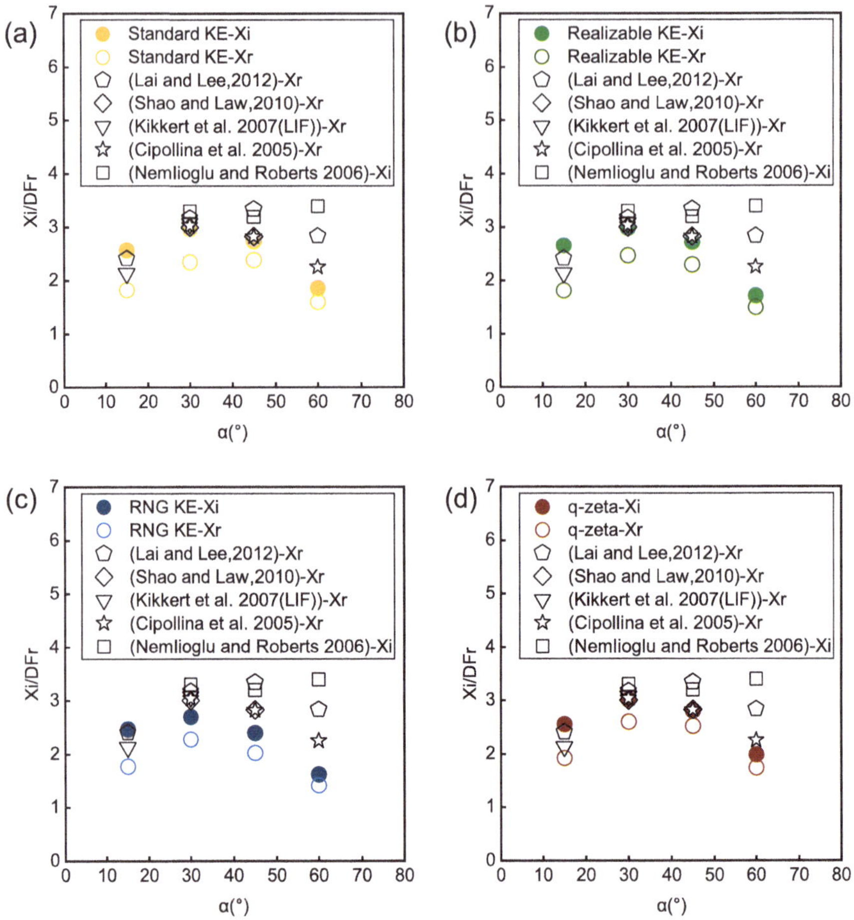

Fig. 9 Normalized location of the return point (X_r/DFr) and impact point (X_i/DFr)plotted against the discharge angle: **a** Standard $k-\varepsilon$ **b** Realizable $k-\varepsilon$ **c** RNG $k-\varepsilon$ **d** q-zeta

from the experimental study [41] exhibited the highest value in comparison with the return point [10, 28]. In future studies, these two points should be addressed and compared separately, as they contain valuable information for outfall design considerations.

Notably, the results from four turbulence models in this study under-predicted compared with previous experimental studies, especially for the discharge angles of 45° and 60°. A similar pattern was observed by Zhang et al. [51]. In their numerical study of the simulation of the 45° discharge angle, the RANS model exhibited an under-predicted trend compared to the experimental study. This under-prediction may be overcome by employing other numerical simulation methods such as large-eddy simulation (LES). However, given the higher computational demands and time constraints associated with LES, RANS models remain indispensable in engineering applications. Further detailed studies are still needed to enhance the numerical modelling capabilities of RANS models.

Among these four turbulence models, the RNG model exhibited the least accurate results compared with the other three models, and the $q-\zeta$ model had the best performance. The difference between the standard $k-\varepsilon$ and realizable $k-\varepsilon$ results was insignificant.

3.2.3 The Dilution Information at the Centerline Peak and Return/Impact Point

Geometrical information provides valuable insights for engineers to predict the physical range of inclined dense jets. Additionally, concentration information at specific locations is of great significance because of its direct relevance to environmental factors, particularly concentration levels. The important points for the inclined dense jet near fields are the dilution information at the centerline peak and impact point, which can be used to assess potential environmental risks and design effective pollution control measures.

The normalized dilution at the centerline peak (S_m/Fr) from the four turbulence models is plotted against the discharge angle (α) in Fig. 10. The dilution at the centerline peak exhibited an increasing trend as the discharge angle increased from 15° to 60°. These results are consistent with most experimental studies, except for one by Shao and Law [46] and a numerical simulation by Gildeh et al. [17], where the 30° discharge angle exhibited a higher dilution than the 45° angle.

As previously mentioned, it is necessary to conduct separate analyses of the return and impact points. Hence, the dilution characteristics at these two points were analyzed separately. The dilution for both points was normalized using the Froude

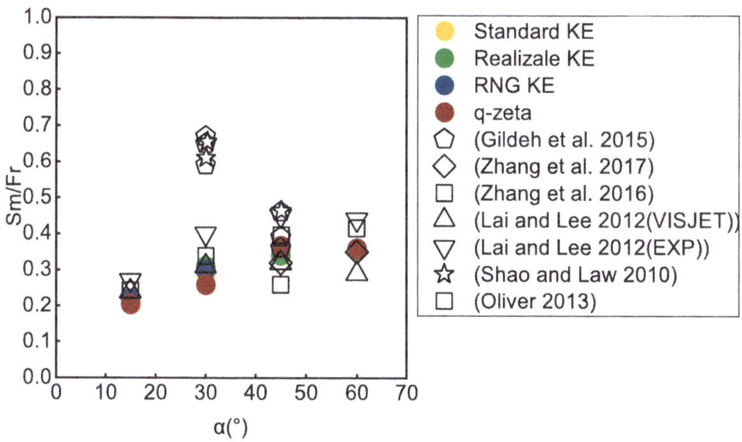

Fig. 10 Centerline peak dilution compared to previous studies

number and plotted against the discharge angle (α), as depicted in Figs. 11 and 12, respectively. In Fig. 11, the return point dilution (S_r/Fr) was compared with previous studies [28, 51, 16, 36] and the results showed notable conformity with the outcomes of commercial tools such as Corjet, UM3, JETLET [36] and other RANS simulations from Zhang et al. [51]. The results obtained from the Large Eddy Simulation (LES), as reported by Zhang et al. [51], demonstrate the closest alignment with the experimental findings of Lai and Lee [28].

In Fig. 12, a comparison is made between the impact point dilution ($\frac{S_i}{Fr}$) and the previous experimental study [31], as well as numerical simulations [33, 51]. The results indicate that the numerical simulations consistently underestimated the dilution compared to the experimental study. However, the four turbulence models

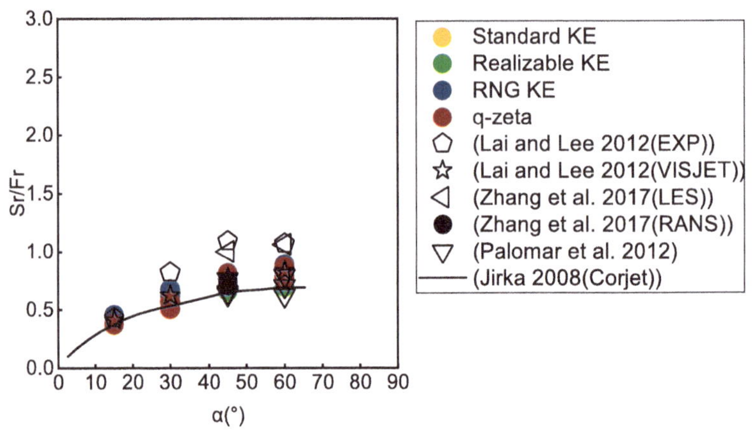

Fig. 11 Return point dilution plotted against discharge angle

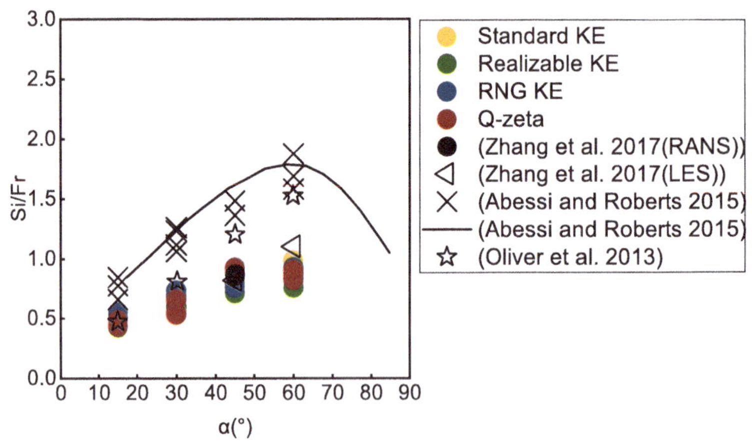

Fig. 12 Impact point dilution plotted against discharge angle

exhibit notable concordance with the numerical simulations presented by Zhang et al. [51] when the discharge angles were 45° and 60° and have good agreement with the Corjet result from Oliver et al. [33] when the discharge angle was 15° and 30°.

Notably, it is worth highlighting that the dilution at both the return and impact points increased as the discharge angle increased. This trend has also been observed in experimental studies. Moreover, the discharge angle of 60° resulted in the highest dilution levels at both the return and impact points. This observation aligns with the research conducted by Roberts et al. [40], who recommended a 60° discharge angle as optimal in the outfall design.

3.3 Cross-Sectional Contration Profile

The previous sections focused on local information at specific points, which may not fully capture the detailed development of the inclined dense jet during its movement. To gain a comprehensive understanding of the jet behavior, it is essential to investigate the concentration fields. The mixing characteristics of an inclined dense jet can be analyzed at different locations using its concentration profiles. Figure 13a shows one of the simulations of an inclined dense jet from the centerline of the computational domain. As evident from the Fig. 13a, the jet ascends with the discharge angle until its highest point and then descends back to the bottom. Throughout this dynamic process, the jet undergoes a transition from jet-like to plume-like, with the occurrence of detraining and mixing during the descending phase [28].

Three concentration profiles were chosen at three different locations along the jet flow (white dashed line in Fig. 13a). On each crossing section, the nondimensional concentration C/C_m is plotted against the dimensionless normalized coordinate r/b_c, where C is the location concentration at the cross-section. C_m is the maximum concentration of the cross-section, r is the radial distance, b_c is the radius defined by $\frac{C}{C_{\max}} = 1/e$.

Figure 13b presents three representative cross-sectional profiles of concentration fields from four different turbulence models that are compared with the results of [28]. Near the nozzle area, the concentration profiles had a Gaussian distribution, which was defined as the "initial stage" of the jet ($s/D = 10$, where s is the streamwise distance and D is the nozzle diameter). During this stage, the jet was dominated by the initial momentum. When it was close to the terminal rise height ($s/D = 30$), in the inner (lower) half of the jet, there was a trend of departure from the Gaussian distribution, whereas the outer (upper) half exhibited a Gaussian distribution. This stage is called the "developed stage," in which the profile is nearly stabilized. After the terminal rise height, at the $s/D = 50$, the upper half of the jet continued exhibiting a Gaussian distribution, whereas the lower half deviated from the Gaussian pattern. This departure from the Gaussian distribution suggested the occurrence of the detrainment in the lower half of the jet. In other words, more brine is peeled from the core flow in this region. Comparing the four turbulence models, the RNG model performed poorly at $s/D = 30$. The same inferior performance of the RNG

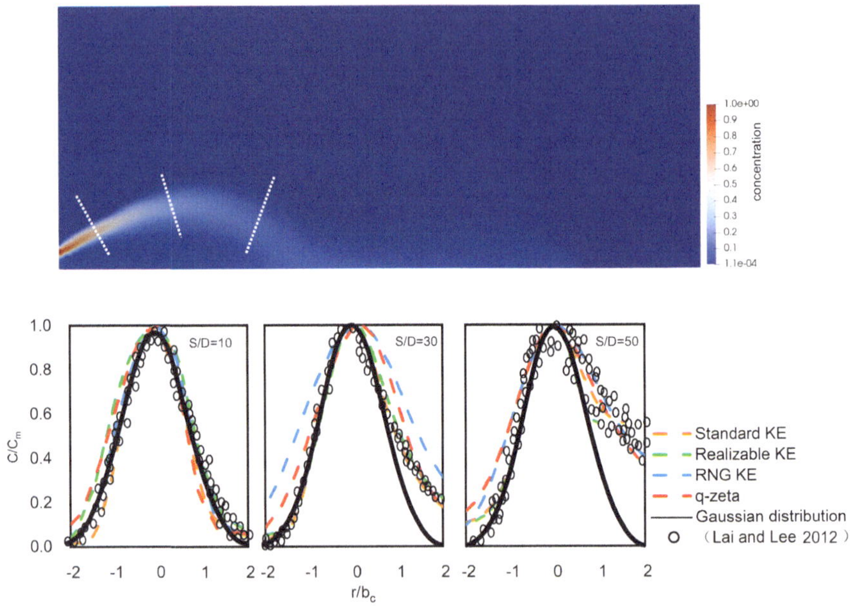

Fig. 13 Simulation result of concentration profile ($\alpha = 30°, Fr = 14.2$): **a** simulation results of realizable $k - \varepsilon$; **b** cross-sectional profiles for $s/D = 10, 30, 50$, respectively

model was found in a study by Gildeh et al. [17]. The other three models exhibited satisfactory performance.

3.4 *Centerline Maximum Velocity Decay*

As the jet was discharged, the flow properties (concentration and velocity) along the centerline gradually changed. It is important to understand this variation along the centerline to provide more information regarding the dissipation of the fluid momentum. In the case of vertical buoyant jets, Chen and Rodi [8] found that the dimensionless centerline dilution (C_m/C_0) along the jet centerline can be represented by a power-law relationship with the Froude number (Fr) and nondimensional vertical elevation (y):

$$F \cdot C_m/C_0 = a(y/D/Fr)^b \tag{33}$$

where C_m is the maximum profile concentration, C_0 is the initial concentration, y is the elevation from the nozzle. a and b are constants related to the flow type. Shao and Law [46] and Gildeh et al. [17] found a similar relationship for the normalized maximum centerline velocity (U_m/U_0). In this study, the same method is adopted.

The variations in the maximum centerline velocity (U_m/U_0)with the non-dimensional downstream distance ($s/D/Fr$) are presented. Due to the limited experimental study, only the 30° and 45° discharge angles were selected and are presented in Figs. 14 and 15, respectively. All figures are plotted on a logarithmic scale. It can be observed that there is a relationship exists between U_m/U_0 and $s/D/Fr$ as Eq. 33. The equations for 30° and 45° discharge angles are $F \cdot U_m/U_0 = 5.237(s/D/F)^{-1.093}$ and $F \cdot U_m/U_0 = 4.831(s/D/F)^{-1.141}$, respectively. The two equations were close to those of Shao and Law [46], which indicated that the results from the turbulence models were accurate. Among the four turbulence models considered in this study, the RNG $k - \varepsilon$ model exhibited a relatively poor performance. In comparison, the other three models exhibited only minor differences in their predictive capabilities.

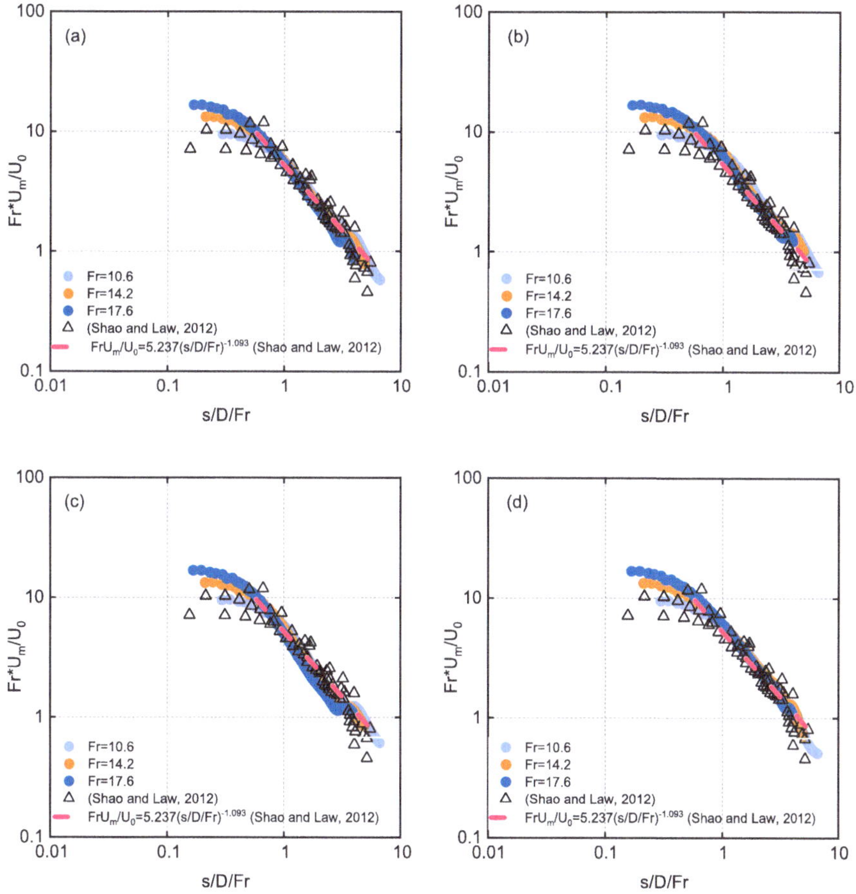

Fig. 14 Comparison of nondimensionalized centerline maximum velocity magnitude decay with 30° discharge angle by four turbulence models: **a** Standard $k - \varepsilon$; **b** Realizable $k - \varepsilon$; **c** RNG $k - \varepsilon$; **d** q-zeta

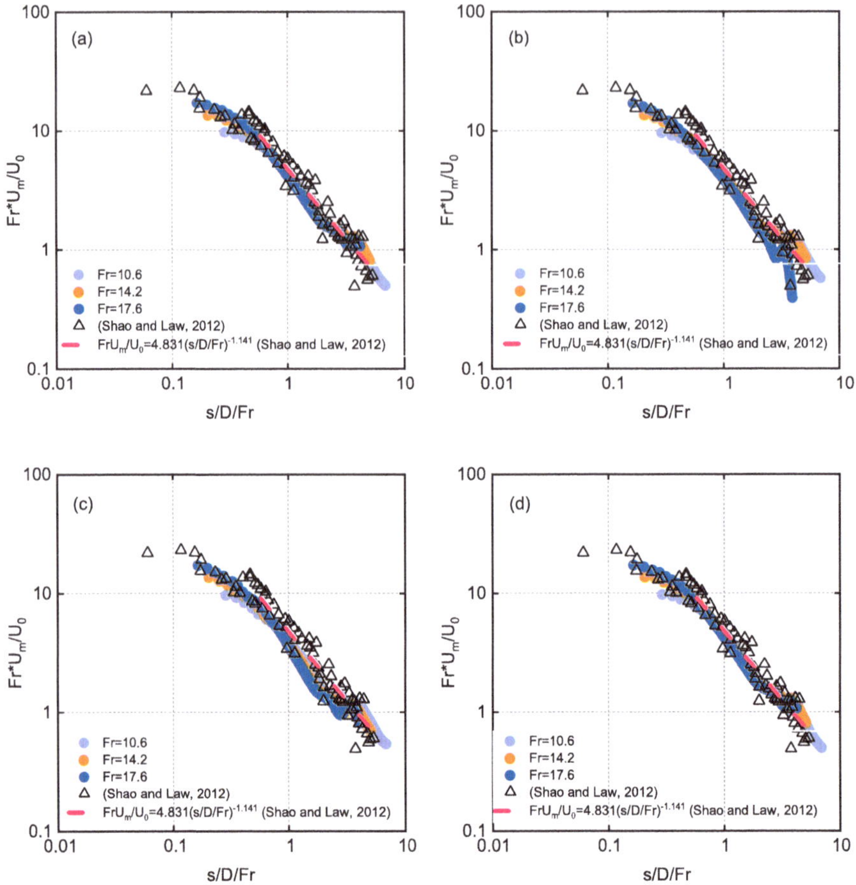

Fig. 15 Comparison of nondimensionalized centerline maximum velocity magnitude decay with 45° discharge angle by four turbulence models: **a** Standard $k-\varepsilon$; **b** Realizable $k-\varepsilon$; **c** RNG $k-\varepsilon$; **d** q-zeta

3.5 Jet Spreading Width

The jet moves and diffuses into the surrounding environment during the movement. Understanding the jet spreading width is critical for engineering design and for assessing the impact area of the jet. Based on the concentration profile analysis, each section of the inclined dense jet had a Gaussian profile for concentration or velocity, especially in the upper half of the jet profile. To comprehensively evaluate the four turbulence models, we analyzed the jet spreading width at 30° and 45° angles. The width of the jet was obtained from the width of the e^{-1} ratio of the maximum concentration or the velocity value of the centerline. All jet width values derived from the concentration (bc) and velocity (bu) profiles are plotted in Fig. 16. All results were consistent with those of previous experimental studies in the range

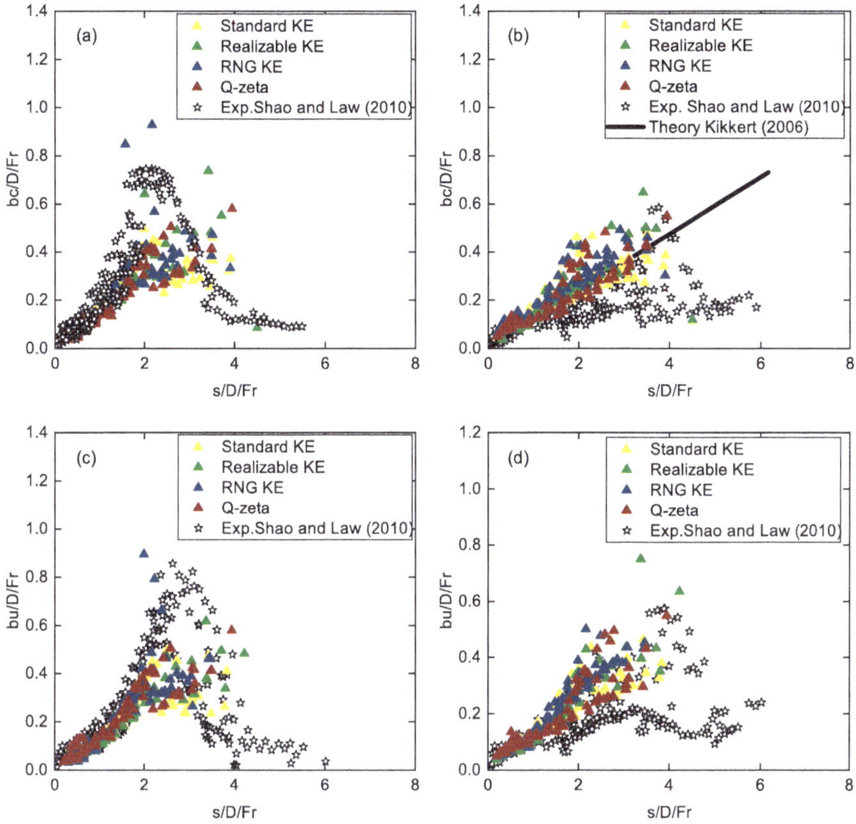

Fig. 16 Variation of the spread width of the concentration (bc) and velocity (bu) along the trajectory for $\alpha = 30°$ and $45°$. (**a**) lower bc; (**b**) upper bc; (**c**) lower bu; (**d**) upper bu

of $s/D/Fr < 4$ [46]. The upper width of the spreading concentration is consistent with the analytical solution from Kikkert [15]. The upper and lower halves of the data increased along the trajectory, and the upper half was less scattered, which was consistent with the experimental results. This indicates that more mixing occurred in the lower part of the jet. The results of the RNG $k - \varepsilon$ turbulence models were more scattered than those of other models.

4 Conclusion and Recommendations

This study presented a comprehensive comparison and evaluation of four $k - \varepsilon$ type RANS turbulence models, namely standard $k - \varepsilon$, realizable $k - \varepsilon$, RNG $k - \varepsilon$, and $q - \zeta$ models, to predict the mixing characteristics of inclined dense jets in stagnant water on the horizontal bottom. The investigation involved four different discharge angles (15, 30, 45, and 60°) with three different Froude numbers using the open-source CFD software OpenFOAM. A comprehensive analysis of the geometric and mixing properties of the jet was conducted, and the results were compared with those of previous experimental and numerical studies.

Among these four models, the RNG $k-\varepsilon$ model exhibited a relatively inferior performance in simulating the geometric characteristics of inclined dense jets. The realizable $k-\varepsilon$, standard $k-\varepsilon$ turbulence, and $q-\zeta$ models provided reliable predictions, including the terminal rise height, centerline peak location, and return and impact points, consistent with previous experimental findings. While the locations of the return point (X_r), impact point (X_i), and centerline peak (X_c, Y_c) of all four-turbulence models were slightly underestimated compared to the experimental data, especially the larger discharge angle ($\geq$45°), but which remained consistent with other numerical simulations. Notably, the RNG $k-\varepsilon$ and $q-\zeta$ turbulence models performed better than the others in predicting dilution at the impact point. All the numerical simulations consistently exhibit a degree of underestimation when compared to the findings of the experimental study, especially for dilution at the impact and return points. Employing advanced simulation methods, such as Large Eddy Simulation (LES) or more complex turbulence models, such as RSM (Reynold stress models (RSM), could be beneficial for improving the predictions of inclined dense jets, especially for larger discharge angles. The dilution observed at both return and impact points showed a similar functional relationship with the discharge angle. In future studies, the properties of the return and impact points should be addressed separately.

With regard to the concentration field, all four turbulence models exhibited a good performance. They accurately predicted the cross-sectional concentration profiles of inclined dense jets with a Gaussian pattern. The standard $k-\varepsilon$ model and $q-\zeta$ models showed proficiency in modelling concentration profiles in the descending part of the jet. In addition, a non-dimensional decay equation ($Fr \cdot U_m/U_0 = a(s/D/Fr)^b$) for the maximum centerline velocity decay in inclined dense jets with 30° and 45° discharge angles was developed. The standard $k-\varepsilon$, realizable $k-\varepsilon$, and $q-\zeta$ turbulence models showed excellent agreement with previous experimental studies, indicating that the decay of the centerline maximum velocity follows a typical jet behavior.

An analysis of the jet spread width revealed that the concentration and velocity had similar developed patterns. All four turbulence models obtained satisfactory predictions for the velocity and concentration fields, which is consistent with previous experimental studies. In addition, the velocity and concentration exhibited slower increases near the upper boundary than near the lower boundary.

In conclusion, the realizable $k-\varepsilon$, standard $k-\varepsilon$, and $q-\zeta$ turbulence models performed well in simulating the geometrical properties and concentration field of inclined dense jets. Geometrical information prediction is limited when the inclined dense jet has a steep discharge angle (60°). Therefore, it is beneficial to use other advanced numerical simulation methods. However, for a mild to moderate discharge angle ($\leq$45°), the realizable $k-\varepsilon$ and $q-\zeta$ models are more recommended owning to their reliable performance.

References

1. M. Alvir, L. Grbčić, A. Sikirica, L. Kranjčević, OpenFOAM-ROMS nested model for coastal flow and outfall assessment. Ocean Eng. **264**, 112535 (2022)
2. B.E. Launder, D.B. Spalding, The numerical computation of turbulent flows, in *Numerical Prediction Of Flow, Heat Transfer, Turbulence and Combustion* (Elsevier, 1983), pp. 96–116
3. R. Bashitialshaaer, M. Larson, K.M. Persson, An experimental investigation on inclined negatively Buoyant jets. Water **4**(3), 720–738 (2012)
4. L.J. Bloomfield, R.C. Kerr, Inclined turbulent fountains. J. Fluid Mech. **451**, 283 (2002)
5. C.C.K. Lai, J.H.W. Lee, Initial mixing of inclined dense jet in perpendicular crossflow. Environ. Fluid Mech. **14**(1), 25–49 (2014). https://doi.org/10.1007/s10652-013-9290-7
6. C.J. Greenshields, OpenFOAM user guide. OpenFOAM Foundation Ltd, Version **3**(1), 47 (2015)
7. C.K. Chu, Mixing of turbulent advected line puffs. *HKU Theses Online (HKUTO)* (1996)
8. C.J. Chen, W. Rodi, Vertical turbulent buoyant jets: a review of experimental data. NASA Sti/Recon Tech. Rep. A **80**, 23073 (1980)
9. P.C.K. Chu, J.H. Lee, V.H. Chu, Spreading of turbulent round jet in coflow. J. Hydraul. Eng. **125**(2), 193–204 (1999). https://doi.org/10.1061/(ASCE)0733-9429(1999)125:2(193)
10. A. Cipollina, A. Brucato, F. Grisafi, S. Nicosia, Bench-scale investigation of inclined dense jets. J. Hydraul. Eng. **131**(11), 1017–1022 (2005). https://doi.org/10.1061/(ASCE)0733-9429(2005)131:11(1017)
11. K. Czaplewski, S. Swierczynski, P. Zwolan, Increasing the reliability of simulation tests in navigation and maneuvering simulators using the k-epsilon model based on the RANS method. Sensors **21**(15), 4995 (2021). https://doi.org/10.3390/s21154995
12. F.J. Millero, A. Poisson, international one-atmosphere equation of state of seawater. Deep Sea Res. Part A. Oceanogr. Res. Pap. **28**(6), 625–29 (1981)
13. F.M. White, J. Majdalani. *Viscous fluid flow*, Vol. 3 (McGraw-Hill, 2006)
14. G. Alfonsi, Reynolds-averaged Navier–Stokes equations for turbulence modeling. Appl. Mech. Rev. **62**(4) (2009)
15. G.A. Kikkert, *Buoyant Jets with Two and Three-Dimensional Trajectories* (2006). https://doi.org/10.26021/1639
16. G.H. Jirka, Improved discharge configurations for brine effluents from desalination plants. J. Hydraul. Eng. **134**(1), 116–20 (2008)
17. H.K. Gildeh, A. Mohammadian, I. Nistor, H. Qiblawey, Numerical modeling of 30° and 45° inclined dense turbulent jets in stationary ambient. Environ. Fluid Mech. **15**(3), 537–562 (2015)
18. H. Jasak, OpenFOAM: Open source CFD in research and industry. Int. J. Nav. Archit. Ocean Eng. **1**(2), 89–94 (2009). https://doi.org/10.2478/IJNAOE-2013-0011
19. S. Hosseini, S.A. Reza, A. Mohammadian, P.J.W. Roberts, O. Abessi, Numerical study on the effect of port orientation on multiple inclined dense jets. J. Mar. Sci. Eng. **10**(5), 590 (2022)
20. A. Issakhov, P. Omarova, Numerical simulation of pollutant dispersion in the residential areas with continuous grass barriers. Int. J. Environ. Sci. Technol. **17**, 525–540 (2020)
21. J. Boussinesq, Theorie de l'ecoulement Tourbillant. Mem. Acad. Sci. **23**:46 (1877)
22. J. Tu, G.H. Yeoh, C. Liu, T. Yao, *Computational fluid dynamics: A practical approach* (Elsevier, 2023)
23. J.D. Anderson, J. Wendt, *Computational Fluid Dynamics*, vol. 206 (Springer, 1995)
24. M. Jiang, A.-K. Law, A.C.H. Lai, Turbulence characteristics of 45° inclined dense jets. Environ. Fluid Mech. **19**(1), 27–54 (2019). https://doi.org/10.1007/s10652-018-9614-8
25. G.H. Jirka, Integral model for turbulent buoyant jets in unbounded stratified flows. Part I: single round jet. Environ. Fluid Mech. **4**(1), 1–56 (2004)
26. E. Jones, M. Qadir, M.T.H. Van Vliet, V. Smakhtin, S.-M. Kang, The state of desalination and brine production: a global outlook. Sci. Total Environ. **657**, 1343–1356 (2019). https://doi.org/10.1016/j.scitotenv.2018.12.076
27. G.A. Kikkert, M.J. Davidson, R.I. Nokes, Inclined negatively buoyant discharges. J. Hydraul. Eng. **133**(5), 545–554 (2007)

28. C.C.K. Lai, J.H.W. Lee, Mixing of inclined dense jets in stationary ambient. J. Hydro-Environ. Res. **6**(1), 9–28 (2012)
29. M.A. Zeitoun, W.F. McIlhenny, Conceptual designs of outfall systems for desalination plants. Offshore Technol. Conf. (1971)
30. M.M. Gibson, A.A. Dafa'Alla, Two-equation model for turbulent wall flow. AIAA J. **33**(8), 1514–1518. https://doi.org/10.2514/3.12691
31. O. Abessi, P.J.W. Roberts, Effect of nozzle orientation on dense jets in stagnant environments. J. Hydraul. Eng. **141**(8), 06015009 (2015)
32. O. Reynolds, IV. On the dynamical theory of incompressible viscous fluids and the determination of the criterion. Philos. Trans. R. Soc. Lond. (A.) **186**, 123–164 (1895)
33. C.J. Oliver, M.J. Davidson, R.I. Nokes, Removing the boundary influence on negatively buoyant jets. Environ. Fluid Mech. **13**(6), 625–648 (2013). https://doi.org/10.1007/s10652-013-9278-3
34. P. Vafeiadou, I. Papakonstantis, G. Christodoulou, Numerical simulation of inclined negatively buoyant jets, in *The 9th International Conference on Environmental Science and Technology* (2005), pp. 1–3
35. P. Moin, K. Mahesh, Direct numerical simulation: A tool in turbulence research. Annu. Rev. Fluid Mech. **30**(1), 539–578 (1998). https://doi.org/10.1146/annurev.fluid.30.1.539
36. P. Palomar, J.L. Lara, I.J. Losada, M. Rodrigo, A. Alvárez, Near field brine discharge modelling part 1: Analysis of commercial tools. Desalination **290** (March), 14–27 (2012). https://doi.org/10.1016/j.desal.2011.11.037
37. I.G. Papakonstantis, G.C. Christodoulou, P.N. Papanicolaou, Inclined negatively buoyant jets 1: geometrical characteristics. J. Hydraul. Res. **49**(1), 3–12 (2011)
38. I.G. Papakonstantis, G.C. Christodoulou, P.N. Papanicolaou, Inclined negatively buoyant jets 2: concentration measurements. J. Hydraul. Res. **49**(1), 13–22 (2011)
39. A.B. Pincince, E.J. List, Disposal of brine into an estuary. J. (Water Pollution Control Federation) **45**(11), 2335–2344 (1973)
40. P.J.W. Roberts, A. Ferrier, G. Daviero, Mixing in inclined dense jets. J. Hydraul. Eng. **123**(8), 693–699 (1997)
41. S. Nemlioglu, P.J.W. Roberts, Experiments on dense jets using three-dimensional laser-induced fluorescence (3DLIF), in *4th International conference on marine waste water disposal and marine environment* (2006), pp. 1–10
42. S. Rodriguez, *Applied computational fluid dynamics and turbulence modelling: Practical tools, tips and techniques* (Springer Nature, 2019)
43. S.B. Pope, A more general effective-viscosity hypothesis. J. Fluid Mech. **72**(2), 331–340 (1975). https://doi.org/10.1017/S0022112075003382
44. S.S. Tong, K.D. Stolzenbach, Submerged discharges of dense effluent. MIT report, no. 243 (1979). https://repository.tudelft.nl/islandora/object/uuid%3A397fe811-d155-430c-83ad-735073f07dc6
45. F.G. Schmitt, About Boussinesq's turbulent viscosity hypothesis: historical remarks and a direct evaluation of its validity. Comptes Rendus Mécanique **335**(9), 617–627 (2007). https://doi.org/10.1016/j.crme.2007.08.004
46. D. Shao, A.-K. Law, Mixing and boundary interactions of 30 and 45 inclined dense jets. Environ. Fluid Mech. **10**, 521–553 (2010)
47. V. Yakhot, L.M. Smith, The renormalization group expansion and derivation of turbulence models **7**(1):35–61 (1992). https://doi.org/10.1007/bf01060210
48. X. Wang, A. Mohammadian, *Numerical Simulations of 15-Degree Inclined Dense Jets in Stagnate Water Over a Sloped Bottom* (Springer, 2022), pp. 63–74
49. X. Yan, A. Mohammadian, X. Chen, Numerical modeling of inclined plane jets in a linearly stratified environment. Alex. Eng. J. **59**(3), 1857–1867 (2020)
50. S. Zhang, B. Jiang, A.-K. Law, B. Zhao, Large eddy simulations of 45 inclined dense jets. Environ. Fluid Mech. **16**, 101–121 (2016)

51. S. Zhang, A.-K. Law, M. Jiang, Large eddy simulations of 45° and 60° inclined dense jets with bottom impact. J. Hydro-Environ. Res. **15**, 54–66 (2017). https://doi.org/10.1016/j.jher.2017.02.001
52. W. Zhao, R. Li, H. Li, Y. Zhang, S. Qiu, Numerical analysis of fluid dynamics and thermodynamics in a Stirling engine. Appl. Therm. Eng. **189**, 116727 (2021)

The manufacturer's authorised representative in the EU is Springer Nature Customer Service Centre GmbH, Europaplatz 3, 69115 Heidelberg, Germany. If you have any concerns regarding our products, please contact ProductSafety@springernature.com

Printed and bound by CPI Group (UK) Ltd, Croydon, CR0 4YY
30/09/2026
02225304-0002